Herbert Kaufman

Time, Chance, and Organizations

Natural Selection in a Perilous Environment

TIME, CHANCE, AND ORGANIZATIONS

TIME, CHANCE, AND ORGANIZATIONS

Natural Selection in a Perilous Environment

Herbert Kaufman

Chatham House Publishers, Inc.
Chatham, New Jersey

TIME, CHANCE, AND ORGANIZATIONS
Natural Selection in a Perilous Environment

Chatham House Publishers, Inc.
Box One, Chatham, New Jersey 07928

Copyright © 1985 by Herbert Kaufman

Publisher: Edward Artinian
Production supervisor: Linda Sabol
Jacket and cover design: Quentin Fiore
Composition: Chatham Composer
Printing and binding: Hamilton Printing Company

Library of Congress Cataloging in Publication Data

Kaufman, Herbert, 1922-
  Time, chance, and organizations.

  Bibliography: p.
  Includes indexes.
  1. Organizational change. I. Title.
HD58.8.K38   1985   302.3'5   85-17477
ISBN 0-934540-40-3
ISBN 9-934540-39-X (pbk.)

Manufactured in the United States of America
10   9   8   7   6   5   4   3   2   1

*. . . the race is not to the swift, nor the battle to
the strong, nor bread to the wise, nor riches to
the intelligent, nor favor to the men of skill; but
time and chance happen to them all.*

Ecclesiastes 9:11

Contents

Acknowledgments

The provenance of this book was a luncheon in 1953 with Richard D. Schwartz, then a member of the Department of Sociology at Yale University, and Donald T. Campbell, then a visiting professor at Yale. Their discussion of social evolution, at a time when I was already engrossed in what turned out to be a lifelong preoccupation with organization theory, set me on a path to which I always seemed to return, no matter how many others I tried.

It was not until 1981, however, when the Russell Sage Foundation appointed me visiting scholar for 1981-82, that I began work on this volume. For the opportunity, and for financial assistance in 1983 that enabled me to finish the manuscript, I am indebted to the foundation and to its president, Marshall Robinson. I am also grateful to all my colleagues at the foundation during 1981-82 — a group so large that individual acknowledgments are infeasible — for reading and commenting on early drafts of the manuscript, and to Madge Spitaleri for typing them. They were patient and kind, and I thank them all heartily.

In addition, the Brookings Institution provided me with an office and logistical support from the autumn of 1982 to the present. To Brookings, to all those colleagues in its Governmental Studies Program — again, too large a group for me to name every one — who provided comments, criticisms, and suggestions, to the superb library staff, and to Joan Milan, Julie J. Bailes, and Judith H. Newman, who prepared the manuscript on the word processor, I should like to express my profound gratitude.

James W. Fesler, Hugh Heclo, and Richard D. Schwartz took exceptional pains to give me both encouragement and tough, constructive criticism, for which I am most appreciative.

This probably would have been a better book had I taken all the counsel I was offered. But my commitment to the logic of my argument as I saw it obliged me to pursue it in my own way. I therefore surrendered the right to spread the blame for its faults. Let not the innocent suffer for my sins.

An Introduction
I Hope You Will Read

I tend to skip over the introductory material in many books, so it is with humility that I ask you to take the time to go through this one. I think it will make a difference to you.

I share your feeling that the authors of books ought to be able to put everything they want to say in the body of their volumes, obviating the need for prefaces and forewords and separate introductions, but many of us are overcome by the need to say something to our readers before the readers get to the substantive discussion. The reasons are not always convincing. Sometimes we yield to the temptation to explain how we came to write the book, though why anyone who has not yet read it should care is something of a mystery; if such history belongs anywhere, it ought to come at the end of the book rather than as front matter. Sometimes we feel obliged to explain the plan of the book, which suggests that the logic of the presentation would otherwise be indiscernible—a confession that ought to drive off a wary reader. Occasionally we use the opportunity to apologize for what the reader is about to read, pointing out what we would do differently if we were starting over; this practice must make many people wonder why the time and energy that went into the apology weren't devoted instead to remedying the deficiencies. Now and then we review all the other ways in which the book might have been organized, presumably to assure prospective readers that they are about to get the best of the lot. Much of the time, I suspect we are simply abiding by the well-known principle of debate, preaching, and exposition: Tell 'em what you're gonna tell 'em, tell 'em, tell 'em what you told 'em; the author of this precept clearly believed no audience would get the point the first time it was presented.

But this introduction is different. I don't want to do any of these things. My objective is to save you time and trouble by telling you in advance what this book is *not* about. You see, some commentators on the manuscript might anticipate something quite different from what it is, in which case they would be rather put out by the discrepancy. I don't want to be guilty of false advertising or deceptive marketing. So I'm putting my cards on the table right at the start.

For example, some readers apparently may think they are going to be treated to a rigorous, elegant, definitive model of the dynamics of populations of organizations. They will be disappointed to discover that the approach is literary, loosely drawn in places, metaphorical, and speculative. For all that, I believe it has many of the qualities we seek in theoretical explanations. It starts with a specific question, posits a dynamics to answer the question, traces out the implications of the dynamics, and proposes steps to explore the usefulness and test the validity of this approach. Though it is a conjecture rather than a body of empirical data or a tightly cast geometric-style proof, conjecture of this kind can be a fruitful part of scientific inquiry. But if you are among those who grow impatient with it, you would be well-advised not to expose yourself to what follows.

And if you are among those looking for a systematic review of the literature on organization theory or on the application of the theory of evolution to social units, you too will be disappointed. Indeed, if you seek an exhaustive bibliography on these subjects, you will not find it here. I have not attempted to fit all prior scholarship into my framework, or to organize existing knowledge in categories suggested by my hypothesis and its implications. Rather, the objective of my source notes and incidental excursuses is to indicate that my assumptions and hunches are not utterly fanciful, spun out of imagination or without consideration of their limits and alternatives to them. I employ them chiefly, if not exclusively, to demonstrate that my premises and many of my inferences are drawn from, and consistent with, doctrines and experience widely (though not necessarily universally) accepted as convincing in authoritative circles. This being my goal, I relied quite often on secondary sources, treatises, encyclopedias, textbooks, and other summary treatments of relevant material, many of them designed as introductions aimed at the uninitiated rather than at seasoned professionals. In them, the reader who is interested in the intellectual foundations of my argument will find précis of studies, monographs, and other original works of scholarship, as well as extensive bibliographies that I have tried to tap in this indirect fashion. When references to original sources seemed appropriate or necessary, I did not hesitate to include them. But if you do not find references to much of the professional literature you expected to appear among the citations in a book of this kind, the reason is probably that their contribution comes through intermediate sources. I have appended to this introduction a short list of works whose command of relevant subjects and literature was particularly helpful to me and strikes me as particularly impressive. Taken together, they provide a comprehensive overview of the wellsprings of much of my argument. They are also superior bases from which to launch further explorations of these subjects, should you be so inclined.

At the same time, I should put such explorers on notice that my argument led me beyond the area of specialization in which I normally operate, and I found myself wandering in fields in which I am an amateur. Anyone better versed in those fields may well discover that I have not invoked the most salient, authoritative, up-to-date works or that my interpretations of those works are occasionally idiosyncratic. Nevertheless, I believe I have not gone so far off the mark as to invalidate the case put forth here. I will be gratified if this volume induces specialists to rectify the weaknesses and improve the formulations in the specialities to which I have not done justice. Meanwhile, I caution those who decide to pursue the line of investigation indicated by my conjecture to take my documentation mainly as a springboard to independent searches of their own. It is not the last word, but the first; not a finish line, but a point of departure; not proof, but illustration; not the history of an idea, but a linkage with earlier and current thought.

It is not a foray into sociobiology, either. Sociobiology, as I understand it, is concerned with the way biological evolution produced in organisms striving for their own survival the self-restraint, cooperation, and altruism frequently manifested when they live in groups. It is, in other words, a quest for the biological origins of social behavior. Obviously its findings impinge on organization theory by illuminating the origins and development of organizations and the genetic factors in organizational structure and behavior, so any comprehensive theory of organizations will eventually have to incorporate some of the findings of sociobiology. In this book, however, my focus is on variation and natural selection among *organizations*, not organisms, and while I am compelled to make some assumptions about selected aspects of human behavior in order to explain the phenomena I deal with, I am not obliged to seek out the genetic origins of that behavior. At this time, sociobiology and organizational evolution may be studied separately and, for convenience, probably ought to be, even though they may one day be brought together. For that reason, sociobiology hardly appears in my discussion, and readers looking for a treatise on that subject should go no further.

Neither should readers anticipate a radical departure from prevailing doctrines and beliefs. The hypothesis advanced here grows directly out of existing systems of thought. If it survives tests of its validity, it would not overturn everything that we have learned and that we believe about human organizations. To the contrary, it builds on them.

That is not to say it is identical with them. It adopts a different perspective, which in turn alters some familiar perceptions. And it forces a head-on confrontation with some questions not central to much of our conventional lore. The logical implications of the hypothesis therefore diverge from traditional precepts, and in nontrivial ways.

But it springs from the same roots they do, it shares a great deal with them, and it relies on many of the same data. Though it adds some new ingredients, it is an incremental variation on themes that have been around for a long time and on ideas about, and studies of, organizational population ecology that

4

have appeared more recently. It does not fly in the face of everything that went before. It does not require abandonment of understanding and beliefs built over generations of experience and thought. Though some of its tenets may take some getting used to, they are in the mainstream of recent developments in organizational studies.

I can't blame you if you're now wondering why in the world anyone *would* be tempted to read the book. I'm not impartial, of course, but I think it contains some novel and interesting thoughts. In particular, I have labored to follow out the implications of many ideas about organizations to which students of the subject generally subscribe, and they led me to conclusions about the characteristics of the world of organizations that are not common currency at this time. Admittedly, the inferences may be wrong. I may have misread the findings and prevailing judgments in the field, or my reasoning may be faulty; that's why I present them as hypotheses. But I have tried hard to be accurate and fair in my characterization of what I take to be existing knowledge and doctrine, I have taken pains to identify those of my positions that are controversial, I have indicated where statements of contrary positions may be found, and I have striven to maintain logical consistency. I believe, therefore, that my argument is on firm ground. And since it has brought me to a view of the world of organizations a little different from the generally accepted ones, I hope it will arouse the curiosity and perhaps win the concurrence of those who pay attention to this sort of thing. For if it is right, it raises questions about the efficacy of some standard organizational strategies and some public policies affecting organizations. Moreover, its validity is not merely a matter of faith; it is testable, and I have described how the tests might be conducted.

Indeed, even if the hypothesis is eventually refuted by tests, it may be worth reading now and it may repay the costs of the testing. Science, after all, is a systematized process of trial and error; it is full of dry holes and blind alleys. The successes are without doubt more gratifying and valuable and exciting than the failures, but the failures are useful too. If they do nothing else, they eliminate some of the many possible explanations of phenomena we labor to understand, thereby narrowing the search to more fruitful avenues. They sometimes stimulate other toilers in the same vineyard to strike out on a line of exploration, or to reformulate a theory, in a fashion that might not have occurred to them and that might prove productive. Also, the data gathered to test a hypothesis that is ultimately rejected may be helpful to other researchers or to policy makers wrestling with altogether different problems; the unanticipated and unintended effects of speculation and research may be extensive. Hence my suggestion that you should not be put off by the substantial possibility that what follows is wrong.

Well, now you know the risks of proceeding further in this volume. I hope you will decide to take the plunge nevertheless, and that the effort rewards your tolerance and your patience.

Addendum: On General Sources

Donald T. Campbell's work in social evolution has been immensely helpful to me for years. In particular, four of his papers (which provide a splendid and up-to-date bibliography on this subject as well as brilliant analysis) are "Variation and Selective Retention in Socio-Cultural Evolution," in *Social Change in Developing Areas: A Reinterpretation of Evolutionary Theory*, ed. Herbert R. Barringer, George I. Blanksten, and Raymond W. Mack (Cambridge: Schenkman, 1965), 19-48; "On the Conflicts Between Biological and Social Evolution and Between Psychology and Moral Tradition," *American Psychologist* 30, no. 12 (December 1975): 1103-26; "Reprise," *American Psychologist* 31, no. 5 (May 1976): 381-84; and "The Two Distinct Routes Beyond Kin Selection to Ultrasociality: Implications for the Humanities and Social Sciences," in *The Nature of Prosocial Development: Theories and Strategies,* ed. Diane Bridgeman (New York: Academic Press, 1983), 11-41.

For excellent surveys of both traditional and recent organization theory, and for the linkage between organization theory and evolutionary theory, I relied heavily on Howard E. Aldrich, *Organizations and Environments* (Englewood Cliffs, N.J.: Prentice Hall, 1979); W. Richard Scott, *Organizations, Rational,*

Natural, and Open Systems (Englewood Cliffs, N.J.: Prentice-Hall, 1981); and John R. Kimberly, Robert H. Miles, and Associates, *The Organizational Life Cycle* (San Francisco: Jossey-Bass, 1980). Other recent major resources are Paul C. Nystrom and William H. Starbuck, eds., *Handbook of Organizational Design,* vol. 1, *Adapting Organizations to Their Environments,* and vol. 2, *Remodeling Organizations and Their Environments* (New York: Oxford University Press, 1981); and Jeffrey Pfeffer, *Organizations and Organization Theory* (Marshfield, Mass.: Pitman, 1982). Michael T. Hannan and John Freeman, "The Population Ecology of Organizations," *American Journal of Sociology* 82, no. 5 (March 1977): 929-64, provide a first-rate discussion of the literature and concepts of relations between organizations and their environments, as well as a challenging and intriguing thesis further developed in their "Niche Width and the Dynamics of Organizational Populations," *American Journal of Sociology* 88, no. 6 (May 1983): 1116-45.

Particularly stimulating parallels between economic and biological theory were drawn in Edgar S. Dunn, Jr., *Economic and Social Development: A Process of Social Learning* (Baltimore: Johns Hopkins University Press, 1971); Jack Hirshleifer, "Economics from a Biological Viewpoint," *Journal of Law and Economics* 20, no. 1 (April 1977): 1-52; Kenneth E. Boulding, *Ecodynamics: A New Theory of Societal Evolution* (Beverly Hills, Calif.: Sage, 1978); and Richard R. Nelson and Sidney J. Winter, *An Evolutionary Theory of Economic Change* (Cambridge: Harvard University Press, 1982).

For the biological elements of my argument, I depended almost entirely on the works cited in the notes. Anyone seeking an introductory overview of that area, however, will probably find G. Ledyard Stebbins, *Darwin to DNA, Molecules to Humanity* (New York: Freeman, 1982), extremely useful and will appreciate the bibliography on pages 457-67.

Q. Why Do Organizations Die?
A. Because Their Engines Stop.

Death comes eventually to the vast preponderance of living things, no matter how benign their environment may be. Something inside them limits their life span. The mechanism is shrouded in mystery, and the differences in the life spans of different species are particularly difficult to explain. Still, whatever the mechanism may be, it is certainly effective.

Yet some living things may not be subject to this built-in limit on their longevity. For example, if an organism reproduces by dividing in two, you cannot tell which is the original and which is the copy. The ambiguity persists each time division takes place. Thus, even if individual replicas have a limited existence, others indistinguishable from the original endure. Consequently it does not die in the conventional sense.

Moreover, certain biological colonies seem never to age because constituent units are continuously replaced, keeping the average age of each colony constant; presumably, then, each colony will go on indefinitely unless destroyed by external forces. Indeed, certain reptiles and fishes appear to be capable of unlimited growth and therefore able to survive indefinitely.[1]

But they are the exceptions. Most species have limited life spans and their members die after their allotted time no matter what they do.

Logically, what is exceptional among organisms ought to be the rule for organizations. That is, if organizations are not destroyed by external forces, there is no self-evident reason why they should not be immortal. Like biological colonies, they need not age. They are open systems that replenish their material and replace their personnel by drawing upon their environment, a process that should enable them to continue as long as needed material and personnel are present in the environment. Organizations would seem to be "naturally" immortal; the demise of organizations is what requires explanation.[2]

There is not much doubt that organizations do, in significant numbers, cease to exist. We know that business failures in the thousands occur every year.[3] We know that government agencies, such as those that flourished in the days of the New Deal (for example, the National Recovery Administration, the Works Progress Administration, the Public Works Administration, the Federal Emergency Relief Administration, and the Civilian Conservation Corps) and World War II (such as the Office of Price Administration, the War Production Board, the War Relocation Authority, the War Shipping Administration, the War Manpower Commission, and the National War Labor Board) are no longer around.[4] Indeed, whole civilizations have virtually disappeared.[5] Small wonder that eminent students of organizations, such as Chester I. Barnard, concluded that long-lived organizations are atypical and that most organizations disintegrate after comparatively short periods of existence.[6] To be sure, there is some evidence that long life may not be quite as unusual as is commonly assumed;[7] the weight of the evidence and of informed opinion, however, is on the other side. Without a good deal more data and longevity figures of a systematic kind we cannot be sure what the death rates and the longevity figures of various types of organizations are, but there is good reason to surmise that they die in larger numbers and live shorter lives than the immortality hypothesis would lead one to expect.

Indicia of "Organization"

But what is it that dies? The essence of organization is so subtle, so elusive, that nobody has ever proposed a definition enjoying widespread acceptance. Indeed, one of the classic works of organization theory opened with a denial of the usefulness of trying to define it:

It is easier, and probably more useful, to give examples of formal organization than to define the term. The United States Steel Corporation is a formal organization; so is the Red Cross, the corner grocery
store, the New York State Highway Department. The latter organization is, of course, part of a larger one—the New York State government. But for present purposes we need not trouble ourselves about
the precise boundaries to be drawn around an organization or the exact distinction between an "organization" and a "nonorganization." We
are dealing with empirical phenomena, and the world has an uncomfortable way of not permitting itself to be fitted into clean classifications.[8]

Unfortunately, if you are going to speak of organizational
longevity and death, the need to identify some criterion of organizational existence is inescapable. You must be able to distinguish, however imperfectly, between organization and nonorganization in order to determine approximately when an
organization came into existence and when it ceased to exist.
The boundaries do not have to be precise to be useful, nor do
the distinctions have to be exact. After all, no hard-and-fast
line separates living from nonliving matter, animal life from
plant life, humankind from other animals, or for that matter,
red from yellow; yet we manage to decide roughly where the
borders are and when they are crossed. Similarly, although some
cases will be troublesome, I believe that most estimates of the
time of formation and dissolution of most organizations—estimates, that is, of the transition from states of nonorganization
to organization and vice versa—are likely to fall within a few
years of each other.

For, although human organizations comprise myriad complex processes and relationships that collectively give them their
form and character, it is not necessary to examine all the processes and relationships to determine whether they constitute
an organization. One activity, common to all organizations,
is a rough measure of an organization's existence. It is the deliberate demarcation of the organization's boundaries. I mean
by "boundary" the criteria employed to distinguish one set of
people from all others (though boundaries can be defined for
groups of objects, for territories, and for bodies of knowledge
and ideas as well). If a set of people does nothing purposefully

to establish who is in the set and who is not, then we can be virtually certain that they will not maintain any of the other processes and relationships characteristic of organizations. Conversely, if we can detect planned steps to admit, identify, and expel members, the group will surely exhibit the other features of organizations.

That is not to say boundary demarcation is what *makes* an aggregate into an organization, any more than common uniforms are what turn a group into an army or a team. Rather, my argument is that the conscious effort by a collection of individuals to set up a boundary around itself is a rough, reliable indicator that the aggregation possesses a number of other traits that most of us would recognize as the traits of organizations.[9] For the moment, we need not be concerned with what those other defining traits are. I will merely assume that if a group has them it will surely demarcate its boundaries, and if it demarcates its boundaries it will almost surely have them.

The activities that establish the boundaries thus distinguish organizations from nonorganizations. For example, all the bald men in the world constitute an identifiable group, but they do not take measures to admit people to their ranks or strip unwanted members of their membership. They are a bounded set but they are not an organization. The same is true of the people walking on a particular street at a given moment. Conceivably such groups might take steps to control or at least mark entry into their midst, and to eject previously admitted members or at least declare them nonmembers. At that point they would become organizations. Until then they are distinguishable but unorganized sets of individuals.

Of course, the stringency of the requirements for entry, the conditions of membership, and the grounds for expulsion vary from organization to organization. Some employ rigorous tests of loyalty, knowledge, ability, training, and character before they accept new members, and signal admission through elaborate rites of induction. Others take in new members with minimal screening and ceremony. Some provide their members with clear marks of identity, such as uniforms, badges or other insignia, or cards; others merely add names to rosters. Some go through involved rituals when members are thrown out, quit, or retire, thereby announcing the actions to all the remaining members and to the general public; others quietly cross a name off a list. Most deal with applicants and members individually (though with varying degrees of care); some confer membership or take it away en masse. Not all memberships are sought or willed; we are born into some (such as the family and the state), while some are thrust on reluctant persons (prison inmates, for example) and boundaries are maintained to keep them in rather than to bar entry. Boundary markers, in short, are not all alike in function, clarity, or rigor.

Consequently, this principle of classification, like all others, grows fuzzy at the margins. People enter and leave some organizations with so little notice, and membership is so casual and nominal, that different observers may disagree about whether the aggregation can be called an organization at all. This very uncertainty, however, suggests that the rigor of the boundary demarcation system may be an index of the strength of an organization's bonds—that is, of the resistance members offer to forces that would separate them from other members, and of the determination and persistence they display in trying to rejoin other members when circumstances do separate them.[10] One would expect organizations with strong bonds to set up very distinct and carefully tended borders, while weakly unified ones would probably be much more nonchalant about entry and claims to membership. Still, whether the borders are rigorously or laxly defined, any group taking any deliberate and discernible measures of this kind will here be regarded as an organization, and the absence of such measures will be taken to mean that an otherwise identifiable group is not an organization. No more is needed to tell the difference between the presence and the absence of organization.

Of course, organizations differ from one another in many respects; they are not a totally homogeneous class. Nevertheless, in the pages that follow, I address myself to their common properties rather than to the distinctive properties of various "species." I do so not because the attributes of individual species are unimportant, but because the answers to the questions addressed in this volume emerge from the broad patterns of the organizational world; the level of analysis in any inquiry must be appropriate to questions posed. For the purposes at hand, the specifics of particular types of organization are not essential, but it *is* vital to know whether a set of persons does or does not constitute an organization. And this, I suggest, requires only examination of their efforts to demarcate and maintain their boundaries. These efforts are the best single indicator of organizational existence.

Organizational "Death"

The start or end of boundary maintenance serves also as a mark of organizational "birth" or "death." This terminology does not imply that organizations literally are born and die in the same sense that organisms do, or that they are forms of life in the same sense that organisms are. Like certain large molecules and coacervate drops of liquid, organizations share a number of attributes with living things, including the ability to maintain their identity for periods of time, but there are enough differences to set them apart.[11] The language used for living things—birth, life, and death—is adopted here only because these terms are more convenient and less convoluted and contrived than more precise substitutes would be. When I say that an organization has been born, I shall mean that a collection of individuals has just established and keeps a boundary around itself, implying all the other processes and relationships people commonly ascribe to organizations; when I say an organization dies, I shall mean that I can no longer discern a maintained boundary; when I say it is alive, I shall mean that the boundary demarcation activities are still being conducted.

Another definitional problem arises from the possibility of organizational changes that are individually so small as to make little difference, yet that accumulate into transformations so extensive that the organizations eventually bear no resemblance to what they were previously.[12] If this occurs, it would seem sensible to conclude that somewhere along the line the original organization disappeared and a new one was formed—

that is, that the original organization died and a new one was born.

On the other hand, who would say that an organization has dissolved because its product is changed, or because it takes a new name, or because many of its members are replaced, or it installs new leaders in office, or its methods are modified, or it adopts new goals, or it moves to a new location, or it enters into relationships with new suppliers or customers, or even if several of these alterations take place at once? In this regard the organization is like a man who changes his name, his trade, his domicile, his appearance, his citizenship, his wife, his style of life, his personality, his religion, and all his organization memberships; nobody would say he died and somebody else was born. Yet if a few discontinuities in his bodily functions occur—loss of heartbeat, blood pressure, breathing, and brain activity—he is pronounced dead, though everything else about him remains the same. Similarly, I shall treat an organization as continuously in existence as long as a group of people uninterruptedly maintain a boundary around themselves no matter what other changes take place, and any interruption (except for momentary breaks) in the maintenance of the boundaries I shall regard as evidence of its death. Even if its functions are continued by other organizations, it will be counted as a fatality. And even if boundaries are eventually reestablished around much the same group, and the group displays many of the same characteristics it had before the interruption, I shall hold it to be a new organization. When the boundaries lapse the organization ceases, as far as I am concerned. No doubt there will be ambiguous cases (just as there are among human beings who linger on the border between life and death), but for the most part I think this indicator will discriminate between organizations that cease to exist and those that undergo substantial transformations yet, like butterflies going through larval and pupal and adult stages, remain the same living individual.

The "Engines" of Organizations

An organization, more or less like a living thing, is a self-sustaining engine.[13] It consumes matter and energy as it conducts activities; the activities, in turn, enable it to acquire more matter and energy; the acquired matter and energy permit continuation of the activities.

Among the activities of any organization is the maintenance of its structure and boundaries. Quantitatively, however, its activities are largely conversions of matter and energy into forms used by individuals and by other organizations, both inside and outside its boundaries, who contribute to it what it needs or wants in the same way that they receive from it what they need or want. The process is circular and, once started, goes on until something disrupts the exchanges or the acts of conversion. If either is interrupted the other soon stops also, and the boundaries disappear. Such interruptions, then, are the immediate causes of organizational death.

The mutual exchange of benefits does not imply that all the benefits are material or that all observers will regard every exchange as equal. The satisfactions that people derive from membership, for example, may induce them to make heavy contributions of their time and energy and material possessions in return for very slight tangible advantages. And organizations as well as individuals will often contribute substantially to other organizations for apparently ethereal rewards, such as humane causes in distant lands.

Nor are the exchanges necessarily simultaneous. People and organizations may contribute in what seem to be expectations of future events, both concrete and intangible, or possibly in gratitude for past events.

In short, the element of exchange is never totally absent from the relations between organizations and their members or from relations among organizations. One can always point to some quid pro quo to explain why relationships endure and why, therefore, organizations maintain their capacity to attract the energy and material things they employ in their conversion activities, even when the contributors get back little of a visible character.

The specific incentives are discussed in greater detail later on. At this point the main thing to note is that organizations are engines of activity that depend on a flow of fuel and supplies to keep going, and that the flow will decline and end if the activities that convert the fuel and supplies into incentives cease. Both must go on or neither can. For many organizations they obviously do stop. The next question, then, is another *why:* Why do they stop? What are the underlying causes of death producing the immediate causes?

NOTES

1. "Some organisms seem to be potentially immortal. Unless an accident puts an end to life, they appear to be fully capable of surviving indefinitely. This faculty has been attributed to certain fishes and reptiles, which appear to be capable of unlimited growth. . . . A distinction can be made between death as a result of internal changes (i.e., aging) and death as a result of some purely external factor, such as an accident." Lawrence Kaplan, "Life-Span," *The New Encyclopaedia Britannica,* 15th ed. (1982).

2. "A noteworthy difference between the social and the biological organism is the certainty of death in the latter. . . . A state or a nation . . . does not need to contemplate its own end, because its units are ceaselessly refreshed." Walter B. Cannon, *The Wisdom of the Body,* rev. ed. (New York: Norton, 1939), 319-20.

3. Dun & Bradstreet, *Business Failure Record, 1982-1983* (New York, 1985); Glenn R. Carroll and Jacques Delacroix, "Organizational Mortality in the Newspaper Industries of Argentina and Ireland: An Ecological Approach," *Administrative Science Quarterly* 27, no. 2 (June 1982): 169-98; and John Freeman and Michael T. Hannan, "Niche Width and the Dynamics of Organizational Populations," *American Journal of Sociology* 88, no. 6 (May 1983): 1116-45.

4. See also Herbert Kaufman, *Are Government Organizations Immortal?* (Washington, D.C.: Brookings Institution, 1976); and Thomas W. Casstevens, "Population Dynamics of Governmental Bureaus," *The UMAP Journal* 5 (1984): 178-99.

5. See, for example, Arnold J. Toynbee, *A Study of History* (New York: Oxford University Press, 1947); Cuthbert T. Patrick, *The Lost Civilization: The Story of The Classic Maya* (New York: Harper & Row, 1974); Rice Odell, ed., *Conservation Foundation Letters* (Washington, D.C.: Conservation Foundation, October and November, 1977.)

6. "Successful cooperation in or by formal organizations is the abnormal, not the normal, condition. What are observed from day to day are the successful survivors among innumerable failures. The organizations

commanding sustained attention, almost all of which are short-lived at best, are the exceptions, not the rule. It may be said correctly that modern civilization is one characterized by the large residue of organizations that are in existence at any given time; but this does not imply that the particular organizations of that time have been or will continue to be in existence long. Similarly, it is recognized that the existence of a population does not necessarily imply longevity, but merely the balancing of constantly recurring deaths by births.

"Thus most cooperation fails in the attempt, or dies in infancy, or is short-lived. In our western civilization only one formal organization, the Roman Catholic Church, claims a substantial age. A few universities, a very few national governments or formally organized nations, are more than two hundred years old. Many municipalities are somewhat older, but few other corporate organizations have existed more than one hundred years. Failure to cooperate, failure of cooperation, failure of organization, disorganization, disintegration, destruction of organization—and reorganization—are characteristic facts of human history." Chester I. Barnard, *The Functions of the Executive* (Cambridge: Harvard University Press, 1938), 5.

See also Bertram M. Gross, *The Managing of Organizations: The Administrative Struggle* (New York: Free Press, 1964), 2:660-65; Arthur L. Stinchcombe, "Social Structure and Organizations," in *Handbook of Organizations,* ed. James G. March (Skokie, Ill.: Rand McNally, 1965), 148-50; David A. Whetten, "Sources, Responses, and Effects of Organizational Decline," in John R. Kimberly, Robert H. Miles, and Associates, *The Organizational Life Cycle: Issues in the Creation, Transformations, and Decline of Organizations* (San Francisco: Jossey-Bass, 1980), 358-59; Carroll and Delacroix, "Organizational Mortality in the Newspaper Industries of Argentina and Ireland," 173-79; and Freeman and Hannan, "Niche Width and the Dynamics of Organizational Populations."

7. Kaufman, *Are Government Organizations Immortal?,* 34.
8. James G. March and Herbert A. Simon, *Organizations* (New York: Wiley, 1958), 1.
9. See, for example, the provisional list of 26 group-properties identified by Robert K. Merton in his *Social Theory and Social Structure,* rev. and enlarged ed. (New York: Free Press, 1957), 310-26. See also Herbert Kaufman, "Why Organizations Behave as They Do: An Outline of a Theory," *Papers Presented at an Interdisciplinary Seminar on Administrative Theory* (Austin: University of Texas, 1961), 40-48.
10. See pages 27, 74, and note 7 in chapter 2.
11. See pages 87-91.
12. Herbert Kaufman, *The Limits of Organizational Change* (University, Ala.: University of Alabama Press, 1975), 114-18.
13. For elucidation of exchange theory applied to organisms, see John Tyler

Bonner, *Cells and Societies* (Princeton: Princeton University Press, 1955), chap. 13; J. H. Rush, *The Dawn of Life* (Garden City, N.Y.: Hanover House, 1955), 167-87; and Mahlon B. Hoagland, *The Roots of Life: A Layman's Guide to Genes, Evolution, and the Ways of Cells* (Boston: Houghton, Mifflin, 1979), chap. 4. For its application to organizations, see Jeffrey Pfeffer and Gerald R. Salancik, *The External Control of Organizations: A Resource Dependence Perspective* (New York: Harper & Row, 1978); Howard E. Aldrich, *Organizations and Environments* (Englewood Cliffs, N.J.: Prentice-Hall, 1979), chap. 11; and W. Richard Scott, *Organizations: Rational, Natural, and Open Systems* (Englewood Cliffs, N.J.: Prentice-Hall, 1981), chaps. 5, 8.

CHAPTER 2

Q. Why Do Their Engines Stop?
A. Usually Because They Develop Resource
Problems.

One possible underlying cause of organization death is that the life spans of organizations, like those of most organisms, are self-limiting and that organizations go through life cycles of youthful vigor, maturity, old age, and death. Just because we have no ready theory of the limiting mechanism, one might argue, we should not reject the possibility; since organizational death is common yet puzzling, the life-cycle proposition cannot be ruled out.[1] Moreover, we are as much in the dark about the mechanisms producing characteristic life spans of organisms as of organizations; if we are willing to accept the existence of such a process, whatever it may be, for the former, why not for the latter in the same fashion?

Indeed, the proponents of this view might push their case even further. There *is* a theoretical reason why organization engines might be expected to stop as they get older; they could get set in their ways, an affliction comparable to hardening of the arteries, and eventually their operations might get so sluggish that death results.[2] It seems to stand to reason.

But this thesis has limited explanatory power. First, it leaves us in the dark about the apparently great variance in the ages at which organizations—even organizations of very similar kinds—die. Some last for human generation after human generation, others that resemble them die young, and probably more die young than grow old.[3] If this impression is right, few organizations would live long enough to go through a full life cycle; the majority would seem to be killed off by other factors.

If the fixed-life-span theory has any validity at all, it would account for a minor proportion of all organizational mortality.

Second, a contradictory hypothesis seems just as appealing. Isn't it plausible that organizations that endure for a long time are flexible, experienced, and wily (or they would have succumbed like their shorter-lived fellows)? One might therefore deduce, as several organization theorists have, that the longer an organization lives, the *greater* its remaining life expectancy and the *fewer* its sclerotic infirmities.[4]

In any event there is something intuitively unpersuasive about a diagnosis of an organization's death that consists merely of a finding that the organization, regardless of its age at the time of death, simply completed its allotted span. It is dubious for any individual death; therefore it is even more dubious for organization deaths collectively.

So I am compelled to reject the idea of fixed life spans as an explanation for most organization deaths.

Two other putative causes undoubtedly do result in the death of organizations: organizational suicide and cataclysmic events. But I believe they account for only an infinitesimal fraction of all the organization deaths in any period.

Suicides occur when the leaders or members of organizations, or both, deliberately and voluntarily terminate them. I think that many ostensible organizational suicides will, on investigation, turn out to be provoked by the imminence of death from other causes; often what looks voluntary is actually done under duress. But many cases of self-destruction are truly voluntary, induced by benefits and rewards to members and/or leaders conditioned on the termination of their organizations; the profitable sale of companies, and mergers in which the combining entities surrender their identity as they are incorporated into a larger one, are illustrations. And there are also instances in which certain foundations, clearly capable of continuing, have been ended because their founders, on principle, decreed that they should shut down after specified periods; the officers of these institutions, being under heavy moral and legal obligations to do so, carried out the wishes of the benefactors. Clear cases of willingly self-inflicted death are not unknown.[5]

Neither are catastrophes that utterly wipe out organizations by dispersing or killing the members. The destruction of Pompeii and of Carthage and the murder of whole communities and families by the Nazis in World War II are cases in point.[6]

Obviously these are extraordinary events. Only a minute proportion of organizations are likely to end in this fashion. Suicidal and catastrophic destructions of organizations may and probably do occur in every era, but they are never likely to cause large numbers of organizations to go out of existence. Self-destruction is especially rare; generally, organizations tend to *resist* termination — they struggle to survive[7] — even if they complete their assigned missions and even when many outside observers are of the opinion that their demise would be a blessing to everyone.[8]

Still, organizations do die and in significant numbers. If limited life span, suicide, and cataclysm do not explain more than a tiny minority of these deaths, what does? The most plausible hypothesis, in my judgment, is that the inflows of energy and other resources necessary for them to keep their activities going, to keep their engines running, dry up. Sometimes the required inflows cease because the resources are exhausted. Sometimes the resources are as available as ever, but the organizations are decreasingly successful in attracting them. Indeed, organizations may be unable even to elicit continued contributions of labor and obedience from their own members. As we gather more information about the demise of organizations I anticipate that such deprivations of essential ingredients will prove the most common cause of death.[9]

Once again we are confronted with a puzzling question: Why do organizations whose existence demonstrates that they must once have had the ability to acquire needed ingredients lose that ability? What deprives them of the ingredients whose lack shuts down the engines of their lives?

NOTES

1. The life-cycle theory is alive and flourishing; see John R. Kimberly, Robert H. Miles, and Associates, *The Organizational Life Cycle: Issues in the Creation, Transformation, and Decline of Organizations* (San Francisco: Jossey-Bass, 1980). See also Marver H. Bernstein, "The Life Cycle of Regulatory Commissions," in his *Regulating Business by Independent Commission* (Princeton: Princeton University Press, 1955), chap. 3; Anthony Downs, "The Life Cycle of Bureaus," in his *Inside Bureaucracy* (Boston: Little, Brown, 1967), chap. 2; and Anthony Downs, "Neighborhood Life Cycles," in his *Neighborhoods and Urban Development* (Washington, D.C.: Brookings Institution, 1981). In addition, cyclical theories have been advanced by a number of students of history; see Oscar Handlin, *Truth in History* (Cambridge: Harvard University Press, 1979), 90-92.

2. See pages 76-77 and note 36 in chapter 3; Herbert Kaufman, *The Limits of Organizational Change* (University, Ala.: University of Alabama Press, 1975), chap. 1. David A. Whetten's "Sources, Responses, and Effects of Organizational Decline," in Kimberly, Miles et al., *The Organizational Life Cycle,* also describes organizational atrophy and the literature treating it (355-58), though decline does not necessarily lead to death.

3. See note 6 in chapter 1.

4. "The older a bureau is the less likely it is to die"; Downs, *Inside Bureaucracy,* 20. "The older an organization is, the more likely it is to continue to exist through any specified further time period"; Samuel P. Huntington, *Political Order in Changing Societies* (New Haven: Yale University Press, 1968), 13. See also Kaufman, *The Limits of Organizational Change,* 99-100; and Whetten, "Organizational Decline," 358-59.

5. For example, "Julius Rosenwald declared that he 'was not in sympathy with this policy of perpetual endowment' and set up the Rosenwald Fund 'with the understanding that the entire fund in the hands of the Board, both income and principal, be expended within twenty-five years of the time of my death.' The Fund came to an end in June, 1948"; F. Emerson Andrews, *Philanthropic Giving* (New York: Russell Sage Foundation, 1950), 100. For illustrations of the self-termination of independent business establishments by sale or merger, mostly voluntarily and mostly entailing loss of identity, see Gertrude G. Schroeder, *The Growth of Major Steel Companies, 1900-1950,* Johns Hopkins University Studies in Historical and Political Science, Series 70, no. 2 (Baltimore, 1952), especially chap. 2 and Appendix tables 13-15.

A different form of self-inflicted organizational death occurs when organizations restrict their membership to a fixed group of people involved in specific events in the past. For example, the veterans of the Union armies in the Civil War formed the Grand Army of the Republic in 1866, which attained a membership of over 400,000 within twenty years. Since later generations were not eligible to join, the organization shrank as members died. It disappeared when the last one died in 1956. Thomas H. Johnson, *The Oxford Companion to American History* (New York: Oxford University Press, 1966), 341. Similarly, the organizations of alumni who graduate in the same year from schools and colleges dwindle until they vanish. Standards of eligibility for membership that prevent admission of new members both by birth into an organization and by recruitment assure the organization's self-destruction.

For the sake of completeness, I note also the extraordinary possibility of mass suicide by all of an organization's members. It has happened! A group of Zealots, a Jewish sect holding the fortress of Masada, now in Israel, took their own lives rather than be taken by the Romans in A.D. 73; Yigael Yadin, *Masada: Herod's Fortress and the Zealots' Last Stand* (New York: Random House, 1966), 11-12, 232-37. In 1978 another sect vanished when more than 900 members of the People's Temple of the Disciples of Christ took poison at their commune in Guyana on the command of their leader; Jon Nordheimer, *New York Times,* 26 November 1978.

6. See also Hans Zinsser, *Rats, Lice and History* (Boston: Little, Brown, 1935), chaps. 7, 8; and William H. McNeill, *Plagues and Peoples* (New York: Doubleday, 1976).

7. Bertram M. Gross, *The Managing of Organizations: The Administrative Struggle* (New York: Free Press, 1964), 2:658-59. Herbert A. Simon, Donald W. Smithburg, and Victor A. Thompson also discuss what they term "the survival drive" of organizations in *Public Administration* (New York: Knopf, 1950), at 402, in the course of two chapters (18 and 19) devoted to "the struggle for existence."

The struggle for organizational existence is illustrated by the tendency of leaders to seek out new, viable goals for their organizations when the original ones are either attained or prove unattainable. For example, the National Foundation, which fights birth defects and arthritis, was originally the National Foundation for Infantile Paralysis, but changed its name and target after poliomyelitis was nearly eliminated. Similarly, the American Lung Association was once the National Tuberculosis Association, but broadened its mission as the incidence and dangers of the disease that had been its raison d'être declined. More generally, see the discussion of "goal succession," and the literature therein cited, in David L. Sills, "Voluntary Associations: Sociological Aspects," *International Encyclopedia of the Social Sciences* (New York: Macmillan and Free Press, 1968), 16:370-72; the same author's *The Volunteers: Means and Ends in a National Organization* (New York: Free Press, 1957), chap. 9; and Charles Perrow, "Organizational Goals," *International Encyclopedia of the Social Sciences,* 11:305-11. See also Erwin W. Bard's examination of the Port of New York Authority's striking changes in program emphases when its attempts to rationalize and integrate railroad operations in the port bore little fruit; *The Port of New York Authority* (New York: Columbia University Press, 1942).

8. Charges of continuation beyond their period of usefulness and justification are frequently leveled against government organizations in general. For example, "a governmental unit can continue for many years after its utility has passed, or its form of organization or program have become obsolete"; Luther Gulick, "Notes on the Theory of Organization," in *Papers on the Science of Administration,* ed. Luther Gulick and L. Urwick (New York: Institute of Public Administration, 1937), 43. "The moment government undertakes anything, it becomes entrenched and permanent. . . . A government activity, a government installation, and government employment become immediately built into the political process itself"; Peter F. Drucker, *The Age of Discontinuity: Guidelines to Our Changing Society* (New York: Harper & Row, 1968), 226. During the latter years of the 1970s there was a flurry of interest in "sunset legislation," statutes intended to eliminate public agencies that could not justify their existence to legislative bodies; see Dan R. Price, *Sunset Legislation in the United States* (Austin, Texas: State Bar of Texas, 1977, processed). One of the grounds for the movement was the premise that "agencies were outliving their usefulness" (Price, 11), were becoming "practically immortal" (U.S. Senate, *Sunset Act of 1977, Hearings before the Subcommittee on Intergovernmental Relations of the Senate Committee on Governmental Affairs,* 95th Cong. 1st sess., 1977, 69) and that "bureaucratic and political pitfalls have made it almost impossible to eliminate programs or agencies once they're on the statute books" (ibid., 68).

But continuation of organizations beyond the time when many observers expect (or demand) their death is evidently not confined to government agencies. Peter Drucker, in the work cited earlier in this note, declared that "of all our institutions, business is the only one that society will let disappear" (237). The federal government's measures to save the Lockheed Aircraft Corporation from bankruptcy in 1971 (Public Law 92-70, 9 August 1971 [85 Stat. 178]; *National Journal,* 7 August 1971, 1676) and the Chrysler Corporation from the same fate in 1980 (Public Law 96-185, 7 January 1980 [93 Stat. 1324]) suggest that even business firms may not give up the ghost if they wield a great deal of political influence when it looks as though their time has come.

Organizations of all kinds, it would appear, fight to survive rather than dying cheerfully or taking their own lives.

9. The interlock of organization and environment has been the subject of considerable attention in recent years; for example, Gross, *The Managing of Organizations: The Administrative Struggle,* vol. 1, chaps. 17, 18; Paul R. Lawrence and Jay W. Lorsch, *Organization and Environ-*

ment (Cambridge: Harvard University Press, 1967); Ephraim Yuchtman and Stanley E. Seashore, "A System Resource Approach to Organizational Effectiveness," *American Sociological Review* 32, no. 6 (December 1967): 891-903; Neil W. Chamberlain, *Enterprise and Environment: The Firm in Time and Place* (New York: McGraw-Hill, 1968); Jeffrey Pfeffer and Gerald R. Salancik, *The External Control of Organizations: A Resource Dependence Perspective* (New York: Harper & Row, 1978); Marshall W. Meyer and Associates, *Environments and Organizations* (San Francisco: Jossey-Bass, 1978); Howard E. Aldrich, *Organizations and Environments* (Englewood Cliffs, N.J.: Prentice-Hall, 1979); Paul Nystrom and William H. Starbuck, eds., *Handbook of Organizational Design,* vol. 1, *Adapting Organizations to Their Environments,* and vol. 2, *Remodeling Organizations and their Environments* (New York: Oxford University Press, 1981); and W. Richard Scott, *Organizations: Rational, Natural, and Open Systems* (Englewood Cliffs, N.J.: Prentice-Hall, 1981). But the interlock's potential *lethality* for organizations, though recognized explicitly or tacitly, has not ordinarily been the major emphasis of such treatments, as it is in Glenn R. Carroll and Jacques Delacroix, "Organizational Mortality in the Newspaper Industries of Argentina and Ireland: An Ecological Approach," *Administrative Science Quarterly* 27, no. 2 (June 1982): 169-98; John Freeman and Michael T. Hannan, "Niche Width and the Dynamics of Organizational Populations, *American Journal of Sociology* 88, no. 6 (May 1983): 1116-45; and in this volume.

CHAPTER 3

Q. Why Do They Develop Resource
Problems?
A. Because Their Environment Is Volatile
and Adjusting to It Is Not Easy.

My starting point is the premise that most of the organizations that die are done in by the interaction of two factors. One is the incessant change, the turbulence, of their environment. The other is their difficulty in adjusting to this volatility. The combination causes them resource problems that I believe are the principal explanations of organizational demise. Were their setting less dynamic or their adjustability greater, I would have to look elsewhere for explanations. Such is not the case. Their built-in properties and the evidently inherent characteristics of their surroundings seem to me to point inescapably to this conclusion.

Why the Organizational Environment Is Volatile[1]

Those features of the world in which organizations are imbedded that bear most heavily on their lot are the outcomes of human intercourse and activity. Few things in the world are more ephemeral.

The physical world, too, has its effects, as we have seen. But on the scale of human lifetimes, most physical changes are so gradual as to be almost imperceptible. Big, sudden ones are unusual enough to be major news.

THE SOCIAL SETTING OF ORGANIZATIONS

What humankind *does* to the physical world, however, in just the eyeblink of time that we call human history, has drawn even the slow and stable elements of geology and biology into the time-scale of human affairs. We have in a few centuries stripped away forests, exhausted soils, depleted supplies of minerals, and fouled air and water that nature needed eons to produce. Not all organizations have been affected equally by these exploitations, but the impact on many has been profound, and the repercussions have doubtless touched many more in ways of which they are scarcely, if at all, aware. Organizational resources dwindle and disappear because organizations use them up at such a fearful rate.[2]

All the same, change in the physical world is slow and deliberate compared to the mutability of social, economic, and political institutions and practices. Fads, fashions, and styles arise, often suddenly; they take hold and spread, often quickly; they peak and decline, often in a short time; and they disappear, often without leaving a residue.[3] In almost every facet of life — nutrition, dress, the arts, sports, recreation, science, standards of beauty and status and elegance, sex, childrearing, education, cooperation and competition, among others — new preferences, attitudes, opinions, and behavior patterns supplant old ones continuously (and frequently swing back to older forms). Even values of a fundamental kind can be expected to shift in time.[4] What is called usury and condemned in one era is called interest and approved in another.[5] Gambling is decried and suppressed by one generation, publicly licensed and even governmentally operated in another.[6] Responsibilities once considered individual become social, and public authorities are called on to monitor the quality and safety of products and services, rescue people and organizations from economic and physical hardship, assure safety in the workplace, reduce disparities of economic power, and in other ways intervene in relationships once considered matters of purely personal concern.[7] Toleration of political unorthodoxy waxes and wanes, and old doctrines are reinterpreted until the original words take on meanings that would doubtless astonish those who formulated them.[8] A few broad tenets, most of which appear in similar form among many of the world's major religions, exhibit extraordinary durability, at least as ideals. But most values and norms governing, or expressed in, human conduct change sooner or later.[9] Not many are set in stone.

Changes in natural resources, tastes, values, politics, and public policies are results, in part, of economic development and experience, but they also shape patterns of economic life. So economic life is marked by great variability.[10] Productivity, the distribution of wealth and incomes, and standards of living fluctuate. Even the most pronounced secular trends are averages of cyclical variations, and the effect of the shorter-term oscillations, let alone the longer-term trends, on the organizational world is powerful. It is not only the organizations traditionally defined as "economic" that feel the impact of these ups and downs; so do all the other kinds of groups in every system. All experts are not necessarily of the same mind on the reasons for these variations, but there is not much disagreement about their ubiquity since the dawn of history. These are a virtually universal characteristic of the environment of human organizations.

So are changes in knowledge, both basic and applied, and in beliefs. Technological developments — applied knowledge — obviously strike organizations dramatically, even organizations in societies that remain highly traditional compared to those that generate and embrace technological change. Technology itself has so multiplied the connections and interdependencies in the world that hardly any group anywhere on the planet is immune to the effects of technological advances. From the discovery of fire and the invention of the wheel, through the revolutions in agriculture, food processing, manufacturing, transportation, and communication, to space travel, genetic engineering, and the creation of artificial intelligence, innovations have altered the environment of organizations, and frequently in amazingly short spans of time. Many of these dramatic alterations have proved afflictions as well as blessings to humankind, but that is beside my point. Whether for good or ill, they utterly transformed the conditions of existence for hosts of organizations everywhere. As long as technology does not stand still — that is, even if it retreats instead of advancing — it will agitate the organizational environment in ways too self-evident and numerous to require rehearsal here.[11]

The impact of pure science and other bodies of belief is more subtle but not necessarily less forceful. The ideas of Copernicus, Newton, Darwin, Einstein, Adam Smith, Locke, Marx, Hegel, Keynes, Freud, Gandhi, of the great religious prophets and books, and of innumerable other articulators of theory and precept have changed the lives of people and groups who never heard of them, or who reject their teachings, or who pay them lip service while violating or perverting their doctrines, not to mention those who do subscribe to their principles and try to abide by them. How people think of themselves, their responsibilities to others, and the way to get what they want are shaped by the interplay among these systems of thought and the patterns of action derived from them. Inevitably, therefore, organizations are molded by these forces and by contacts with other organizations holding other views. I am not asserting that history and society are products of ideas alone; ideas clearly depend just as much on history and society as the reverse. All I am saying is that when we try to understand the world of organizations, we must appreciate that the role of abstract ideas and systems of thought should not be overlooked. The appearance of such constructs, their success in winning adherents (both informed and unwitting), their displacement or modification by other concepts, and their competition with one another help make the environment of organizations as inconstant as it is.[12]

In addition, the composition and distribution of the human population in which organizations form keep changing. Not only do hosts of people move every year, but birth and death rates go up and down, longevity patterns therefore shift, age and sex distributions take on new configurations, the size and power of ethnic and religious and regional and political groups rise and decline, social and economic classes develop new relationships, old homogeneities break up and new ones appear, and the anthropological map is rarely the same from one decade to another. And the rates of change are not the same everywhere or over time, either; local and regional patterns as well as national and worldwide ones vary a great deal. The picture changes nonuniformly, capriciously. Thus does demography intensify the variability of the organizational environment.[13]

Yet when all is said and done, the most volatile element in the organizational environment may not be nature, or social structures and preferences, or ideas, or demography. Rather, it may be other organizations.[14] For most organizations are contained by other organizations, contain other organizations, and/ or overlap other organizations (i.e., share with them some members, but not all). If organizations are constantly being born and dying, and if their structure and behavior are not absolutely fixed while they live, it follows that the major component of their environment is continuously in flux. Although the vital statistics on organizations are fragmentary, those we have do give the impression that they form and dissolve and change with some frequency.

ORGANIZATIONS AS ORGANIZATIONAL ENVIRONMENT

Organizational Births and Deaths. A high rate of organizational births is certainly what one would expect on the basis of logic alone. People recognize that they can accomplish much more by dividing labor and cooperating than is possible through isolated, individual toil. Sheer common sense dictates that they should get together whenever possible to join their efforts. Thus, apparently, do many organizations arise.[15]

But portraying organizations as products of human will and intelligence may not capture the whole story. Another view is that organizations are in a sense genetically determined— that in the long run the people who survived the perils of humanity's early existence were those with a genetic tendency toward life in organizations.[16] Another is that we learn to live in groups long before we learn to think logically; the period of prolonged dependence on family and other groups during our formative years conditions us to life within organizations— and to reject solitary life even if we could seriously contemplate it.[17] Whichever view is right—or whatever balance between them is right—they both emphasize the strongly ingrained character of human membership in groups. Organizations are not necessarily creations of the intellect alone.[18]

Some of our most powerfully bonded organizations, families and extended kinship groups, could conceivably have formed without deliberate design.[19] After all, many nonhuman animals form such collectivities, some of them quite complex, but few people attribute these groups to calculated purpose on the part of the members even though membership does confer demonstrable benefits on individual participants.[20] Perhaps human associations, though much modified and refined by reason, stem from comparable roots.

The argument need not be confined to blood groups. Societies could conceivably emerge from the same constellation of circumstances. To be sure, some social-contract philosophers have presented their theories as though primeval people contrived social orders to escape from the chaos and dangers of nonsocietal existence. But even these theories were probably put forth more as heuristic devices than as literal descriptions of historical events.[21] One could as plausibly assert that humankind is an artifact of society as that society is an artifact of humankind.[22] Both are apparently products of nature.

Similarly, what organization theorists are pleased to call "informal organizations"—organizations not officially established or sanctioned by the ritually designated and generally recognized authorities of the organizations in which they occur, and not congruent with the official hierarchy of authority—seem to arise without conscious plan or intent. Rather, they emerge from the routine relations between people regularly in contact with one another. Understandings and standardized practices develop so gradually and unconsciously that they become norms before those who are governed by them are aware they have crystallized. Sometimes they supplement and strengthen formal structures; sometimes they clash with them, frustrating and undermining the wishes of the formal authorities. Sometimes they are mobilized by external interests; usually they are spontaneous. They are often strong and enduring. Yet for the most part they are not set up, they simply appear.[23]

Indeed, even organizations whose provenance is the calculated quest for benefits cannot be explained entirely by will and intent. Merely wanting to establish an organization, for any reason, is not ordinarily enough to bring it into existence; the environment must offer the opportunities and provide the support that enable the will and intent to create an organization to come to fruition. Purpose and determination are but two among many factors.[24]

What is more, they are not the most helpful explanatory factors. Often, by looking at other circumstances, we can predict when organizations will form.

> In many cultures, for example, the intersections of heavily traveled routes or the sites of breaks in transportation (requiring changes or renewals of vehicles) almost always become the sites of services for the travelers. And wherever a flow of commerce is prohibited or heavily taxed, smugglers are sure to set up shop. We may not be able to tell exactly who will articulate the goal and initiate the action, but when many necessary conditions converge, we can be confident initiators will emerge. In a sense, human design may be likened to a nearly omnipresent contributing factor that plays its part as soon as other elements set the scene, much the way oxygen is essential to combustion, but rarely starts fires on its own. In most cases, if the other things are there, you can rely on it to make its contribution.[25]

This is not circular reasoning. I am not just saying that organizations appear when conditions are ripe and conditions are ripe when organizations appear. What I suggest is that we can forecast the appearance of certain organizations before anyone has indicated the wish or formulated a plan to set them up, and we can find cases when the goal is clear and the will is strong yet the wished-for organization never takes shape.

In a constantly changing environment, conditions hospitable to the birth of new organizations must occur frequently. Since the appearance of new organizations further alters the environment, it will create opportunities for still other organizations, and the formation of new ones must go on endlessly. The process seems logically to be self-sustaining. For this reason alone one would expect organizational birth rates to be high and continuing.[26]

Together, all the postulated elements contributing to the birth of new organizations—the benefits of organizing and the unplanned thrusts toward collective life—make up a formidable combination of forces.Under the circumstances it would hardly be surprising if new ones formed in profusion, helping to keep their own environment in a state of constant agitation.

By the same token, the deaths of organizations would add to environmental volatility. Every time an organization dissolves, others with which it dealt are affected in some way, at least temporarily, and the people who composed it are set loose for a while in the system. In many cases the effects are highly localized and of short duration, but when large organizations cease to exist, the shock waves through the environment may be of tidal proportions.

There is not much doubt that organizations do die in significant numbers.[27] Whether or not my hypothesis about the *reasons* for these deaths is valid, the reasoning about the *consequences* of the deaths for the organizations' environment stands independently. Whatever makes the engines of organizations stop and thus leads to their demise, the unending disappearance of organizations from the scene contributes to the turbulence of that environment. The births and deaths of organizations roil it constantly, keeping the population of organizations forever changing.

Organizational Activity. Moreover, individual organizations are not utterly changeless, as though they were set in concrete. In the course of time their composition, membership, intakes, outputs, and processes may vary repeatedly. (For reasons to be explained, these variations do not mean that adjustment to environmental volatility is therefore easy.) And since every organization is interlocked with others through overlapping memberships[28] and exchanges of goods and services, each such alteration of structure or behavior impinges on its neighbors and trading partners. Their own actions intensify the inconstancy of their surroundings.

One reason organizations are not static, it seems to me, is that in general they "abhor" uncertainty (in the way that nature is said to "abhor a vacuum").[29] Uncertainty, within their boundaries or in the segments of their environment touching their own activities, apparently sets off a process common to most organizations that continues until the uncertainty is reduced to approximately the level that obtained before the disturbance. The degree of uncertainty that sets the process in motion is probably not the same for all organizations, but all organizations doubtless have some limits to their tolerance. Since uncertainty is never completely eliminated, the process is sure to be triggered here and there in any organization during any interval of time. That irritability maintains the level of uncertainty generally, thereby generates new responses, and thus keeps the process going endlessly. (Presumably the chain reactions could become explosive and destroy a whole population of organizations. Occasionally such revolutions do occur. But dampers of some sort seem to keep things from reaching such a destructive state very often. If my analysis is right, understanding what the dampers are and how they work is a task to be undertaken.)

The process set off by external uncertainty is a tendency toward increased organizational self-containment.[30] One form it takes is incorporation of the source of uncertainty within the organization—that is, expanding the boundaries to include it—thereby making it subject to the norms and controls of the system. It is applied to the natural environment by harnessing and managing its forces—substituting agriculture for harvesting wild crops, animal husbandry for hunting, irrigation for rainfall, flood control for flooding, and temperature control for weather, for example. In the case of uncertainty produced by other organizations, this technique takes the form of absorbing them or joining with them in confederal systems or federations.

When expansion is blocked, another expression of the tendency toward organizational self-containment is reduction of exchanges across boundaries in an effort to satisfy most needs and wants internally—withdrawal from the source of uncertainty, as it were. For example, a firm fearing trouble with external suppliers may decide to manufacture its own components or develop its own service staffs instead of contracting for them. Similarly, nations often stockpile critical substances in order to diminish their dependence on foreign sources. Indeed, *autarky*—national self-sufficiency—though it inevitably lowers living standards and elevates costs within the nation that attempts it, has occasionally been urged, and even sought, as a goal for some countries beset by what the advocates believed to be external dangers.[31] Japan managed to isolate itself from the rest of the world almost totally for a long period.[32] Escape is a common reaction to uncertainties that originate outside.

If the origin of uncertainty is internal, the response is usually a striving toward centralization. The techniques include control by the encompassing organization's leaders of all communication to the members of the vexing suborganizations and groups so that nonconforming thought (and therefore deviant behavior) is prevented; intensifying surveillance to discourage nonconformity by increasing the probability of exposure and punishment; detaching key operations (control of funds, recruitment of personnel, promotion, and rewards, for example) from contained organizations, thus reducing their self-containment and increasing their vulnerability to central direction; and training and indoctrinating all members of the system to respond only to commands from the central leadership and from no other source. (If these fail, then the offending suborganization may be expelled—a contraction of boundaries constituting a withdrawal from internal sources of uncertainty.) These measures never succeed totally, especially since new internal groups form all the time. Furthermore, centralization is self-limiting because central decision-making mechanisms and lines of communication tend to get clogged and sluggish, engendering pressures for decentralization. So the tensions never disappear and the efforts never cease.

Thus, organizations are themselves responsible for much of the variability of their environment. They jiggle their boundaries to increase their immunity from the organizations that contain them and to increase their influence over the organizations they contain. They subject individuals, each of whom belongs to more than one organization, to influences that compete and clash, producing unpredictable effects. They thereby add to the change that buffets organizations without surcease.

A Portrait of the World of Organizations. That is why I concluded some years ago that

> the organizational world bubbles and seethes. Observed for a lengthy interval, the configuration of organizations within it changes like the patterns of a kaleidoscope. Organizations expand, contract, break up, fuse. Some surfaces become thick and opaque, reducing exchanges between their interior contents and the external environment, while others etherealize and permit heavier traffic in one or both directions. Shapes are altered. Some processes are depressed, some intensified. Levels of activity rise and fall. Organizations disintegrate and vanish as others form in droves, and the birth and death rates vary over time and space. Nothing stays constant.[33]

The general social setting of organizations accounts for a great deal of this, the organizations themselves for as much or more.

ADJUSTMENT AND SURVIVAL

It stands to reason that an organization in the grip of absolutely fixed patterns of structure and behavior that could not be changed one iota in any respect, if such a thing were imaginable, would soon be in serious trouble trying to survive in such a dynamic environment.[34] Its modes of exchange with its environment would gradually be disrupted; contacts would break off, outputs would diverge from the demands and expectations of the external world, intake pipelines would be severed, and life-sustaining relations would dwindle to nothing. Without these, the wherewithal to continue to elicit contributions to the activities of the organization from its own members would dry up, as would the essential material for keeping up its activities. In short, it would develop the resource problems that stop its engines, as I said earlier.

But organizations are not frozen in a fixed mold. They do change. We can all think of, or at least imagine, instances in which they find new sources of supply when old ones are cut off, or switch to substitute materials when traditional ones are no longer available, or adopt new procedures and processes as old methods grow obsolete, or modify their products and services to maintain or improve their exchanges with the environment, or alter their structure, or in other ways do things differently in response to new circumstances.[35] You would think, then, that their flexibility, coupled with their evident reluctance to go out of existence, would inspire them to adjust to environmental variability and thus preserve themselves. The turbulence of their surroundings should not be lethal to them if they have the capacity to fit themselves to new conditions. Why then do I hypothesize that resource problems arising from the volatility of the environment will turn out to be the most common cause of organizational death? Why do I contend that they have difficulty adjusting?

Why Adjusting to the Volatile Environment Is Not Easy[36]

Organizational adjustment implies more than just any kind of organizational change. Rather, it suggests organizational change matched to change in the environment in a fashion that compensates for the new conditions and keeps the organization running as well as or better than it did before. It refers to a moving equilibrium, as it were, in which the continuity of the organization is never in question even though the cumulative effect of the changes gradually alters many of its features.

Carefully calculated measures are frequently necessary to achieve such appropriate responses to changing conditions. In organisms such responses may be inherent in a creature's anatomy and physiology, and its reactions need not involve anything we would call thought. In human organizations, although analogous automatic responses may occur, adjustments more frequently entail reasoned assessments of the relevant conditions and of the changes considered most likely to achieve the desired end. Adjustment is a strategic calculation. But it rarely is a simple matter.

To explain why this is so I must peer into the interior of organizations. Up to now I have treated organizations as the proverbial black box—as entities whose behavior could be understood by observing them from the outside. And, for some purposes, treating them in this fashion is entirely appropriate. But when it comes to accounting for the failure of some of them to adjust to the ever-changing environment, there is no avoiding an examination of their inner workings. That is where the explanation lies. It has three parts. In the first place, differences of opinion about whether organizational changes are necessary and what changes should be made usually divide the organization's leaders and their advisers and also the members who concern themselves with such things. People of more or less equal wisdom and virtue and knowledge often end up taking different sides on questions of this kind.

In the second place, the way these decisions are reached in most organizations does not ensure outcomes appropriate to the circumstances. The process of organizational decision making does not *prevent* adequate, and even optimal, decisions. But in general it entails a substantial probability that in many instances the outcomes will be ineffectual and perhaps downright pernicious.

In the third place, the execution of organizational decisions is often far from perfect, so that what is actually done in many cases does not carry out the intent and strategy of the decision makers and sometimes even negates their wishes.

Thus, while adjusting to the volatile environment is the obvious theoretical means by which organizations can keep themselves alive indefinitely, the practical difficulties of adjusting are not so easily overcome.

CONTRADICTORY JUDGMENTS

People's perceptions and values result from many factors, including their experiences outside and inside the organizations to which they belong and their physical and mental endowments. Not even identical twins are exactly alike in all these respects, and the leaders and members of most organizations comprise widely differing individuals, not identical siblings. No matter how intensively the organizations endeavor to reduce these differences, and to implant in each person the same outlooks and values and behavior, the range of difference remains broad.[37] So it is hardly surprising that people often disagree about whether the environment has changed enough to require an organizational change in response, about the extent and direction of advisable change, about the feasibility of competing proposals for change, about the probable consequences of various courses of action, about the desirability of each of the consequences — in short, about the past, the present, and the future.

Thus it happens that the leaders of an organization will be told by some factions that survival requires prompt changes of policy and by other factions to keep the course of the organization steady as it goes. Among those who recommend revisions of structure or behavior, some will propose daring forays into untried fields, others will demand equally drastic shifts backward toward older practices, and some will counsel only incremental departures in one direction or the other. Each will insist that the existence of the organization will be placed in jeopardy by the policies urged by the others. In every quarter will be found advocates of comparable stature, intelligence, character, goodwill, loyalty, and records of achievement.[38]

Differences in perceptions and values are intensified by differences in interest. Usually organizational change means a gain for some people and a loss for others. Status, influence, security, job satisfaction, career prospects, pay, perquisites of office, and vested interests in specific programs and processes and job skills and specialized information are among the stakes affected. That is not to say that change in organizations is a zero-sum game in which the rewards of the gainers precisely equal the costs of the losers; the relationships are not that symmetrical, and it is even possible for everyone to gain or lose simultaneously through modifications of structure or procedure, or for the advances and deprivations of each individual to be evenly balanced. As a general rule, however, whenever modifications are introduced, some people come out distinctly ahead and others distinctly behind. Understandably, the winners and losers take different positions on the wisdom of proposed departures from familiar norms.[39]

Personality differences also prompt variations in judgment. For example, some people chafe under stability and welcome

48

change for its own sake; others are uncomfortable with the unfamiliar and thrive under continuity. Some are present-oriented and emphasize quick returns; others are more concerned about the future and recommend long-range planning. Some are optimistic and confident and therefore take risks readily and even eagerly; others are pessimistic and fearful and avoid risk at almost any cost. The need for change, the kinds of change that are possible, and the ratio of returns to dangers will all look very dissimilar to people of contrasting temperaments, who will usually be in opposite corners on questions of organizational change.[40]

For all these reasons, though everyone agrees that organizations must ordinarily adjust to survive, failure to adjust probably accounts for the largest share of organizational deaths. With different people, each from his or her own special vantage point, offering conflicting analyses and recommendations, the optimal (or even the merely adequate) responses to environmental threats to survival are hard to discern. And simply averaging out the recommendations is no guarantee of success; it may sacrifice the strengths of each option and realize the full advantages of none. So figuring out how to adjust is never easy.

DECISION MAKING IN ORGANIZATIONS

But I would go even further. I would say that if you were designing a machine to fashion effective responses to environmental challenges, you probably would not set it up to work the way most human organizations actually work.

I don't mean that they deliver only ineffective decisions, but that the probability of ineffective decisions emerging from the process that obtains in most organizations is higher than it would be in a system engineered for maximum rationality. Human organizations are not like an ideal machine because human beings are so complex and varied that getting them to work together to arrive at decisions is a very uncertain, imperfect art. The best we can do is none too good at producing efficacious responses to the environment.

For instance, if you were trying to design a more rational decision-making process, you doubtless would not give a decisive role to the strongest participant simply for strength, or to the most aggressive for pushiness, or to the most fluent for glibness, or to the loudest, or the most fawning, or to the one most adept at fabricating "data," or to the spokespersons for extremely narrow interests, or to the most doctrinaire, or to the most senior merely for length of service, or to the most self-serving and unprincipled. Yet it is often true in the decision-making councils of organizations that people with these attributes carry the day by virtue of these qualities rather than by the content of their ideas, the wisdom and soundness of their thought.[41] That is not to say people with these attributes never have good ideas and never give wise counsel, but the rationality of decisions is likely to vary widely and unpredictably if personal traits play as large a part in forming decisions as they often do.

Of course, organizations are not deliberately contrived to give greater influence to personality than to analytical capabilities. It often happens, however, that people with great analytical skill and large amounts of information can see the uncertainties of their position and therefore are somewhat hesitant advocates compared to those who see things in oversimplified terms. Fanatics and rigid ideologues often prevail in contests with more openminded and tolerant adversaries, who are inclined to compromise more readily. Scrupulous individuals frequently find themselves at a disadvantage in debates with zealots because they feel obliged to disclose evidence against their position of which the opposition is unaware, while their opponents are unmoved by such meticulousness. Persons who do not fit the dominant organizational stereotype are likely to carry less weight than more traditional types, although the deviants may be free of the mental blinders that limit the conformists' imagination and field of vision. Similarly, timid and modest and shy and inarticulate and unpopular people may have valuable contributions to make to organizational decisions, but they are less likely to have an impact because of their personal characteristics.

Despite all these factors increasing the likelihood that, in the formation of organizational decisions, important considerations will be overlooked or misunderstood and great emphasis will be placed on false premises, effectual responses to environmental variation obviously do occur. Nevertheless, it stands to reason that the chances of responding in ways that hurt instead of helping adjustment must be substantial. This property of organizations is one of the main reasons why adjusting often proves so difficult.

My argument does not assume that evil and folly and ugliness and internal warfare dominate organizations. All I am saying is that the decisions they reach reflect many things besides rational calculations and objective assessments of evidence. This is inevitable when people must come to terms with one another — and even in the most autocratic organizations mutual accommodation is the rule. In these circumstances, organizations could hardly be expected to arrive at effective solutions of their problems with their environment all the time.

Indeed, despite its adverse consequences in some respects, mutual accommodation may be advantageous because it permits people of many kinds, with conflicting interests, to remain together in organizations. Rational decision making is only one of many values served by organizational life.

But typical organizational decision-making processes are far from ideal. They do not assure, and may even impede, efficacious responses to the stresses and strains imposed by the volatility of the environment.

IMPERFECT IMPLEMENTATION

Compounding the problem is slippage between what is decided and what actually gets done by most of the people in an organization. In theory, once a decision is reached by the formally designated decision-making authorities, the behavior of all the people in the organization upon whom the decision bears is directed to translating the words of the authorities into concrete action. In practice the linkage is frequently extremely loose. What leaders aim to have the organization do in response to the environment is often far removed from what happens.[42]

Although the slippage may sometimes be beneficial because decisions that would have been harmful had they been carried out as formulated are transformed into better adjustments, decisions that might have been superb adjustments had they been executed exactly as they were framed could imaginably be turned into disasters by looseness in implementation. Logically, one outcome is as probable as the other. Since I am concerned at this point only with an explanation of the reasons for failures to adjust, I cite only the second possibility here. (But I deal with the other in the next chapter.)

Of course, not every decision falls far short of complete implementation. Though only a few would measure up to *everybody's* standard of full compliance, most are probably comparatively close to the mark. Still, when we examine the causes of destruction of organizations by the environment, I predict that significant among them will be full or partial failure to execute decisions that from all indications would have produced different (and, from the point of view of the organizations, happier) outcomes.

There are several reasons for such slippage.[43] One is that decisions are often ambiguous, and it is not clear what actions are required to put them into operation. After all, decisions forged in controversies among people of varied and possibly clashing perceptions, values, interests, and predispositions will almost inevitably contain obscure and inconsistent provisions; ambiguity is a solvent of difference and a catalyst of consensus. Moreover, instructions come from many sources in organizations, not just one (although each source claims to be acting in the name of the same authorities); since the sources are not necessarily informed about each other's directives, and may in any case be indifferent or possibly hostile to each other, not even the most elaborate clearance procedures eliminate conflicts in these orders. In the end, therefore, subordinates are obliged to develop their own interpretations of what is wanted and expected of them. The results are frequently discrepant with respect to each other and at odds with the expectations and intentions of a majority of decision makers.

Another reason for slippage is that the decision-making process produces some agreements based on unrealistic assumptions about what can be accomplished in the field. On occasion this happens because the decision makers are uninformed or misinformed about the situation in the field.[44] At other times decision makers make knowingly overoptimistic promises in order to satisfy critics or quiet discontent or win support or get contracts.[45] In either case the decisions are not implemented because the rank and file cannot carry them out.

But even if the leaders' decisions are not absurdly impractical, and even if they are not unclear in any important respect, they still may not be executed by the bulk of the organization's members. Instead, the members sometimes go on doing what they did before in the way they grew accustomed to doing it. Some students of organizations refer to this persistence of established practices as inertia.[46] From the point of view of the advocates of change it usually looks like deliberate resistance, if not sabotage.

And sometimes it is. People who calculate that they will be worse off under the proposed changes than they are under the status quo often will not cooperate with them, though their opposition may be covert rather than openly defiant. On the other hand, resistance sometimes stems from deeply implanted behavior patterns and values, and from systemic obstacles, such as sunk costs, the inability to mobilize needed resources, accumulations of official constraints on action, and unofficial and unplanned constraints. Moreover, the introduction of change generally sets in motion limiting factors that restrain the departure from familiar configurations; the threat of innovation arouses the defense of things as they are. The result is organizational action much closer to what prevailed before the attempts to change it than to what the initiators of change envisioned.[47]

A New Puzzle

Trying to adapt to the changing environment thus is loaded with uncertainty for organizations. Reaching a decision on what to do is often a confused and confusing labor because of the differences of opinion about the best way of coping and because the process of deciding does not necessarily give as much weight to rationality as to other qualities. And the actions that follow decisions frequently turn out to bear little relation to the policies they are supposed to carry out. A successful response to an environmental challenge can be a very fortuitous thing.

In view of these difficulties of adjustment and the great volatility of the environment, it would seem that a few years of organizational life are a mighty feat, a few decades a triumph, a few centuries a rarity, a millennium almost an impossibility. Yet long-lived organizations are not so unusual as to occasion wonder. Though they are not profuse, they do occur.[48] How can such longevity be reconciled with the argument I have been advancing here? If effective adjustment is as difficult and uncertain as I have claimed, how do any organizations continue for extended periods?

NOTES

1. Of course, the degree of turbulence and activity is not everywhere the same; Howard E. Aldrich, *Organizations and Environments* (Englewood Cliffs, N.J.: Prentice-Hall, 1979), chap. 3. But I hypothesize that, with a few possible exceptions (see pages 77-79), organizations are subject to at least some of the many kinds of environmental change inventoried here, and that among those that die, environmental variations will prove to be the most common cause of death. Compare with Paul C. Nystrom and William H. Starbuck, *Handbook of Organizational Design*, vol. 1, *Adapting Organizations to Their Environments* (New York: Oxford University Press, 1981).

2. Rice Odell, "History Offers Some Warnings on Environment," *Conservation Foundation Newsletter* (Washington, D.C.), October 1977, and the works cited therein, especially William L. Thomas, Jr., ed., *Man's Role in Changing the Face of the Earth* (Chicago: University of Chicago Press, 1956); and Rice Odell, "Can Technology Avert the Errors of the Past?" *Conservation Foundation Newsletter* (Washington, D.C.), November 1979.

3. For discussions of fads, fashions, and styles, see Herbert G. Blumer, "Fashion," *International Encyclopedia of the Social Sciences* (New York: Macmillan and Free Press, 1968), 5:341-45; and E. H. Gombrich, "Style," ibid., 15:352-61.

4. "Value" is not an unambiguous term. I intend to use the word as it is defined by Clyde Kluckhohn and others, "Values and Value-Orientations in the Theory of Action," in *Toward a General Theory of Action,* ed. Talcott Parsons and Edward A. Shils (Cambridge: Harvard University

Press, 1951), 388-403, but I fear I have sacrificed rigor in the interests of brevity. See also Robin M. Williams, Jr., "The Concept of Values," and Ethel M. Albert, "Value Systems," in *International Encyclopedia of the Social Sciences,* 16:283-91, for a review of the literature and the doctrines on this complex subject.

5. Don Patinkin, "Interest," *International Encyclopedia of the Social Sciences,* 7:472-73.

6. Edward C. Devereux, Jr., "Gambling," *International Encyclopedia of the Social Sciences,* 6:58-61. Many states and local governments in the United States today operate or license gambling enterprises as sources of revenue though most of these same authorities had been committed to suppressing them as socially unacceptable only a generation or two earlier.

7. Yair Aharoni, *The No-Risk Society* (Chatham, N.J.: Chatham House, 1981), chaps. 1-3.

8. See, for example, C. Herman Pritchett, "Interpretation of the Constitution," in his *The American Constitution* (New York: McGraw-Hill, 1959), chap. 4; and Edward Shils, "The Concept and Function of Ideology," *International Encyclopedia of the Social Sciences,* 7:66-76.

9. See Robin M. Williams, Jr., "The Concept of Norms," *International Encyclopedia of the Social Sciences,* 11:204-18.

10. Edward F. Denison, *Why Growth Rates Differ: Postwar Experience in Nine Western Countries* (Washington, D.C.: Brookings Institution, 1967), chap. 19; idem, *Effects of Selected Changes in the Institutional and Human Environment upon Output per Unit of Input,* Brookings Institution, General Series, Reprint 335 (Washington, D.C., 1978); and Simon Kuznets, *Economic Growth of Nations: Total Output and Production Structure* (Cambridge: Harvard University Press, 1971), chap. 1.

11. Lewis Mumford, *Technics and Civilization* (New York: Harcourt Brace, 1934); Emmanuel G. Mesthene, *Technological Change: Its Impact on Man and Society* (Cambridge: Harvard University Press, 1970); Jacques Ellul, *The Technological Society* (New York: Knopf, 1967); and Harvey Brooks, "Technology, Evolution, and Purpose," and Emmanuel G. Mesthene, "How Technology Will Shape the Future," in *Science, Technology, and National Purpose,* ed. Thomas J. Kuehn and Alan L. Porter (Ithaca, N.Y.: Cornell University Press, 1981), 35-80. Leslie A. White, the noted anthropologist, "saw technology as constituting the most powerful set of determinants of cultural systems and went as far as to say that human societies could be thought of as social ways of operating technological systems"; Robert L. Carneiro, in his review of White's life, works, and thought in *International Encyclopedia of the Social Sciences,* Biographical Supplement, 803-7, at 805. This view, in turn, recalls Karl Marx's emphasis on modes of production of material goods as a driving force of history, though one leading authority denies Marx was ever guilty of a mechanistic technological determinism; see

Shlomo Avineri, *The Social and Political Thought of Karl Marx* (New York: Cambridge University Press, 1968), 72-77.

The "new combinations" of means of production and credit (which is to say the organizational skills) that permit the introduction of new methods and products into an economy, heavily emphasized by Joseph A. Schumpeter and attributed by him to the efforts of innovative entrepreneurs (e.g., *The Theory of Economic Development: An Inquiry into Profits, Capital, Credit, Interest, and the Business Cycle* [Cambridge: Harvard University Press, 1934], 74-94, 223-36), may also be considered technological factors agitating the organizational environment. Lewis Mumford, in *The Myth of the Machine: Technics and Human Development* (New York: Harcourt, Brace, and World, 1967), esp. chap. 9, also assigns great weight to the effect of social "megamachines" on the social environment, but takes an extremely pessimistic view of their consequences.

12. Crane Brinton, *The Shaping of the Modern Mind: The Concluding Half of Ideas and Men* (Englewood Cliffs, N.J.: Prentice-Hall, 1950); Max Weber, *The Protestant Ethic and the Spirit of Capitalism* (New York: Scribner's, 1958); Barbara Ward, *Five Ideas That Changed the World* (New York: Norton, 1959); and Thomas S. Kuhn, *The Structure of Scientific Revolutions* (Chicago: University of Chicago Press, 1962), chap. 10. The belief that the idea is an autonomous force in history may be traced back to Georg W. F. Hegel; Oscar Handlin, *Truth in History* (Cambridge: Harvard University Press, 1979), 93-94.

13. The effects of population *growth* are described in National Academy of Sciences, *Rapid Population Growth: Consequences and Policy Implications* (Baltimore: Johns Hopkins University Press, 1971); Lester R. Brown, *In the Human Interest: A Strategy to Stabilize World Population* (New York: Norton, 1974); Colin Clark, *Population Growth and Land Use*, 2d ed. (London: Macmillan, 1977); Ester Boserup, *Population and Technological Change: A Study of Long-Term Trends* (Chicago: University of Chicago Press, 1981); Just Faaland, ed., *Population and the World Economy in the 21st Century* (New York: St. Martin's Press, 1982).

The effects of population *movement* are dealt with in Leszek A. Kosinski and R. Mansell Prothero, eds., *People on the Move: Studies on Internal Migration* (London: Methuen, 1975); James L. Sundquist, *Dispersing Population: What America Can Learn from Europe* (Washington, D.C.: Brookings Institution, 1975); Brian J. L. Barry and Lester P. Silverman, eds., *Population Redistribution and Public Policy* (Washington, D.C.: National Academy of Sciences, 1980); and Mary M. Kritz, Charles B. Keely, and Silvano M. Tomasi, *Global Trends in Migration: Theory and Research on International Population Movements* (New York: Center for Migration Studies of New York, 1981).

Some of the effects of population *composition* are examined in

Samuel Lubell, *The Future of American Politics,* 2d ed. (New York: Doubleday, 1956), chaps. 3, 4, 5; Mark R. Levy and Michael S. Kramer, *The Ethnic Factor: How America's Minorities Decide Elections* (New York: Simon and Schuster, 1972); Cynthia H. Enloe, *Ethnic Conflict and Political Development* (Boston: Little, Brown, 1973); Nathan Glazer and Daniel P. Moynihan, eds., *Ethnicity: Theory and Experience* (Cambridge: Harvard University Press, 1975), esp. 1-26; Martin O. Heisler, ed., "Ethnic Conflict in the World Today," *Annals of the American Academy of Political and Social Science* 433 (September 1977); and Lance Liebman, ed., *Ethnic Relations in America* (Englewood Cliffs, N.J.: Prentice-Hall, 1982).

14. "The web of organizations" is discussed in Peter M. Blau and W. Richard Scott, *Formal Organizations: A Comparative Approach* (San Francisco: Chandler, 1962), 195-99; "the interconnectedness of organizations" in Jeffrey Pfeffer and Gerald R. Salancik, *The External Control of Organizations: A Resource Dependence Perspective* (New York: Harper & Row, 1978), 69-71. See also Aldrich, *Organizations and Environments,* chap. 11; and note 28 in this chapter.

15. "When two men cooperate to roll a stone that neither could have moved alone, the rudiments of administration have appeared;" Herbert A. Simon, Donald W. Smithburg, and Victor A. Thompson, *Public Administration* (New York: Knopf, 1950), 3. Many other students of organizations similarly stress shared purposes as the cause of cooperative action; W. Richard Scott, *Organizations: Rational, Natural, and Open Systems* (Englewood Cliffs, N.J.: Prentice-Hall, 1981), 19-21. For a fuller classification of incentives inducing people to form or join organizations, see James Q. Wilson, *Political Organizations* (New York: Basic Books, 1973), chaps. 2 and 3; Aldrich, *Organizations and Environments,* chap., 7; and, Scott, *Organizations,* chap. 7.

16. Edward O. Wilson, *Sociobiology: The New Synthesis* (Cambridge: Harvard University Press, 1975), pt. 1.

17. Robert M. MacIver, *The Web of Government* (New York: Macmillan, 1947), chap. 2; Melville J. Herskovits, *Man and His Works: The Science of Cultural Anthropology* (New York: Knopf, 1948), chap. 3; and René Dubos, *Man Adapting* (New Haven: Yale University Press, 1965), 8-10.

18. For an interesting discussion of the formation of two kinds of organization, see Jack L. Walker, "The Origins and Maintenance of Interest Groups in America," *American Political Science Review* 77, no. 2 (June 1983): 390-406; and Jacques Delacroix and Glenn R. Carroll, "Organizational Foundings: An Ecological Study of the Newspaper Industries of Argentina and Ireland," *Administrative Science Quarterly* 28, no. 2 (June 1983): 274-91.

19. One must be careful in talking about families and other kinship groups. In different cultures they may take widely differing forms based on quite different relationships among members. They also have many conse-

quences, some of which are called their "functions" or "purposes" by some commentators. Raymond T. Smith, "Family: Comparative Structure," *International Encyclopedia of the Social Sciences* 5:301-13. Moreover, their origins are only matters of speculation and inference; Herskovits, *Man and His Works,* 114-16. Nevertheless, some kind of human group regarded as familial by virtually every informed observer seems to occur almost everywhere, and none of them seem to be planned or consciously contrived by those who study them or compose them; William Graham Sumner and Albert Galloway Keller, *The Science of Society* (New Haven: Yale University Press, 1927), 1:19-20.

Friedrich A. Hayek would probably call each such grouping a "spontaneous order" (for which he adopts the Greek word *kosmos*), as against a deliberately constructed order (or *taxis*). Both are the result of human action, but only the latter is shaped by human design. *Law, Legislation and Liberty,* vol. 1, *Rules and Order* (Chicago: University of Chicago Press, 1973), 20-22 and chap. 2.

20. Wilson, *Sociobiology,* chaps. 18-26; J. P. Scott, Irven DeVore, and V. C. Wynne-Edwards, "Social Behavior, Animal," *International Encyclopedia of the Social Sciences,* 14:342-65.

21. Willmoore Kendall, "Social Contract," *International Encyclopedia of the Social Sciences,* 14:376-81.

22. F. A. Hayek, *The Three Sources of Human Values* (London: London School of Economics and Political Science, 1978), 6-9.

23. F. J. Roethlisberger and William J. Dickson, *Management and the Worker* (Cambridge: Harvard University Press, 1939), chaps. 22, 23, and pp. 449-62; esp. p. 524: "The men had elaborated, spontaneously and quite unconsciously, an intricate social organization around their collective beliefs and sentiments." See also, George C. Homans, *The Human Group* (New York: Harcourt, Brace and World, 1950), chaps. 5, 6.

24. John R. Kimberly, "Initiation, Innovation, and Institutionalization in the Creation Process," in John R. Kimberly and Robert H. Miles et al., *The Organizational Life Cycle: Issues in the Creation, Transformation, and Decline of Organizations* (San Francisco: Jossey-Bass, 1980), 23-26; Andrew H. Van de Ven, "Early Planning, Implementation, and Performance of New Organizations," ibid., 86-87; and Johannes M. Pennings, "Environmental Influences on the Creation Process," ibid., chap. 5.

Indeed, "organizational goals" have proved to be a rather slippery concept; see Scott, *Organizations,* 261-67. Charles Perrow went so far as to argue that "the serious student of organizational goals finds the matter so complex and categories and concepts so interdependent that there is no certainty what should be labeled a goal, where it comes from, how it changes, and what impact it has"; "Organizational Goals," *International Encyclopedia of the Social Sciences,* 11:310. This is not, however, a universally shared opinion; see, for example, Herbert A.

Simon, "On the Concept of Organization Goal," *Administrative Science Quarterly* 9 (1964): 1-22.

25. Herbert Kaufman, "The Natural History of Organizations," *Administration and Society* 7 (August 1975, as corrected at 365 in November 1975): 135-36.

26. "Organizations tend to call forth organizations"; Bernard Berelson and Gary A. Steiner, *Human Behavior: An Inventory of Scientific Findings* (New York: Harcourt, Brace, and World, 1964), 366.

There are many reasons why organizations stimulate – not necessarily intentionally – the appearance of other organizations. Among the stimuli identified by Simon, Smithburg, and Thompson in their *Public Administration* are the quest for coordination and control of specialized units at lower levels (210-13, 272-79), for improving communication and bargaining with other specialists (291-94), and imitativeness (38, 279). John Kenneth Galbraith, *American Capitalism: The Concept of Countervailing Power* (Boston: Houghton Mifflin, 1952), argued that concentrations of economic power in one set of organizations spawn other organizations on "the other side of the market" to check that power (119); thus, "in the ultimate sense, it was the power of the steel industry, not the organizing abilities of John L. Lewis and Philip Murray, that brought the United Steel Workers into being" (121), and "in precise parallel with the labor market, we find the retailer with both a protective and profit incentive to develop countervailing power whenever his supplier is in possession of market power. The practical manifestation of this, over the last half-century, has been the spectacular rise of the food chains, the variety chains, the mail-order houses (now graduated into chain stores), the department-store chains, and the cooperative buying organizations of the surviving independent department and foodstores" (124). Similarly, the creation of armies in one country spurs the creation of armies in its neighbors.

But cooperation as well as competition and conflict among organizations fosters new organizations – transportation firms to carry goods between resource producers and fabricators, and between fabricators and distributors; communication organizations to connect them with one another; banks to finance their operations; construction companies to erect the structures and build the roads required by the other activities; accounting and law firms; lobbying agencies; personnel suppliers; and so forth. The denser the population of organizations, the higher the birth rate seems likely to be; see pages 107-9.

That does not mean the density will never stop increasing; births may be equaled or exceeded by deaths. My point is only that organizational births themselves are one source of the environmental instability that I believe engenders and kills organizations in substantial numbers.

27. See note 3 in chapter 1.

28. Chester I. Barnard, in *The Functions of the Executive* (Cambridge: Har-

vard University Press, 1938), speculated that "including families, businesses of more than one person, various municipal corporations, autonomous or semi-autonomous governments and branches of government, associations, clubs, societies, fraternities, educational institutions, religious groups, etc., the number of formal organizations in the United States is many millions, and it is possible that the number is greater than that of the total population. . . . Moreover, there are in a short period of a day or a week many millions of formal organizations of short duration, a few hours at most, which are not named and are seldom thought of as organizations" (4). "If A, B, C, D, and E constitute a group of five, then subgroups may be made as follows: ten pairs, ten triplets, five groups of four, one of five. If only one person be added to the group of five, the possible subgroups become fifteen pairs, twenty triplets, fifteen groups of four, six groups of five, and one of six" (108-9).

29. This section summarizes the argument presented in Herbert Kaufman, "Why Organizations Behave as They Do: An Outline of a Theory," in *Papers Presented at an Interdisciplinary Seminar on Administrative Theory* (Austin: University of Texas, 1961), 37-72.

30. Simon, Smithburg, and Thompson, *Public Administration,* 266-67.

31. See Charles S. Tippetts, *Autarchy: National Self-Sufficiency,* University of Chicago Press, Public Policy Pamphlets no. 5 (Chicago, 1933), for a critical discussion of several advocates of this policy in recent times; MacIver, *The Web of Government,* 365-67; Herbert Heaton, *Economic History of Europe,* rev. ed. (New York: Harper, 1948), 476, 512, 710, 711; George N. Halm, *Economic Systems: A Comparative Analysis* (New York: Rinehart, 1951), 341-42.

32. Edwin O. Reischauer, *Japan: The Story of a Nation* (New York: Knopf, 1970), 94-95, 109. In the early seventeenth century, Japanese rulers not only closed down all foreign intercourse but forbade Japanese to go abroad and all Japanese residents abroad to return.

33. Kaufman, "The Natural History of Organizations," 143.

34. Barnard, *The Functions of the Executive,* 6; Simon, Smithburg, and Thompson, *Public Administration,* chaps. 18 and 19; Homans, *The Human Group,* chap. 4; Philip Selznick, *Leadership in Administration: A Sociological Interpretation* (New York: Harper & Row, 1957), 5-22, 29-38; W. Warren Haynes and Joseph L. Massie, *Management: Analysis, Concepts, and Cases,* 2d ed. (Englewood Cliffs, N.J.: Prentice-Hall, 1969), 250-54; Bertram M. Gross, *The Managing of Organizations: The Administrative Struggle* (New York: Free Press, 1964), 2:660-72; and Amittai Niv, "Organizational Disintegration: Roots, Processes, and Types," in Kimberly and Miles et al., *The Organizational Life Cycle,* 375-94.

35. That is not to say that all organizational change is deliberate. Much of it is unplanned; see Herbert Kaufman, *The Limits of Organizational Change* (University, Ala.: University of Alabama Press, 1971), 41-44.

And even when it is introduced by design, organizational survival or welfare may not be the primary motive of the initiators; Anthony Downs, *Inside Bureaucracy* (Boston: Little Brown, 1967), 198-203; Kaufman, *The Limits of Organizational Change,* 44-45. But much change is doubtless calculated to increase the prosperity and security of the organizations in which it is effected, which is why much of the literature on management is given over to methods of achieving innovation — e.g., Robert H. Guest, *Organizational Change: The Effect of Successful Leadership* (Homewood, Ill: Dorsey, 1962); Victor A. Thompson, *Bureaucracy and Innovation* (University, Ala.: University of Alabama Press, 1969); Robert T. Golembiewski, *Renewing Organizations: The Laboratory Approach to Planned Change* (Itasca, Ill.: Peacock, 1972); Gerald Zaltman, Robert Duncan, and Jonny Holbek, *Innovations and Organizations* (New York: Wiley, 1973); and Gerald Zaltman and Robert Duncan, *Strategies for Planned Change* (New York: Wiley, 1977).

36. Many obstacles to innovation are set forth in the references listed in note 35 which explain generally why adjustment is not easy. The classic analysis, first published in 1940, is Robert K. Merton, "Bureaucratic Structure and Personality," now chap. 6 of his *Social Theory and Social Structure,* rev. and enlarged ed. (New York: Free Press, 1957). See also Downs, *Inside Bureaucracy,* 158-60; Kaufman, *The Limits of Organizational Change,* chaps. 1, 3; and idem, *The Administrative Behavior of Federal Bureau Chiefs* (Washington, D.C.: Brookings Institution, 1981), chap. 3; Paul S. Goodman et al., *Change in Organizations: New Perspectives on Theory, Research, and Practice* (San Francisco: Jossey-Bass, 1982), chap. 2 (Chris Argyris), chap. 3 (Barry M. Staw), chap. 4 (Clayton P. Alderfer), chap. 8 (Kenwyn K. Smith), and pp. 416-20 (Robert Kahn).

Some commentators have suggested that the ability to adjust tends to approach zero — that is, a potentially lethal level — in some political systems; see, for example, Wallace S. Sayre and Herbert Kaufman, *Governing New York City: Politics in the Metropolis* (New York: Russell Sage Foundation, 1960), 716-20 (but see also the introduction to the paperbound edition, published by Norton in 1965, at xlv-xlvii); James MacGregor Burns, *The Deadlock of Democracy: Four-Party Politics in America* (Englewood Cliffs, N.J.: Prentice-Hall, 1963), esp. 1-7; idem, *Leadership* (New York: Harper & Row, 1978), 413-17; Michel Crozier, *The Stalled Society* (New York: Viking, 1974); and Mancur Olsen, *The Rise and Decline of Nations: Economic Growth, Stagflation, and Social Rigidities* (New Haven: Yale University Press, 1982). More generally, see Robert L. Heilbroner's comments on "the inertia of history" in *The Future as History* (New York: Harper, 1960), 193-97.

37. Simon, Smithburg, and Thompson, *Public Administration,* 68-73, 76-78; and Herbert Kaufman, *The Forest Ranger: A Study in Administrative Behavior* (Baltimore: Johns Hopkins University Press, 1960), 66-72, 75-83, 232-34.

38. See, for example, Leonard R. Sayles and Margaret K. Chandler's treatment of scientists in planning, in their *Managing Large Systems: Organizations for the Future* (New York: Harper & Row, 1971), chap. 3. For the way experts divide over the causes and cures of economic problems, see Craufurd D. Goodwin, ed., *Exhortation and Controls: The Search for a Wage-Price Policy, 1945-71* (Washington, D.C.: Brookings Institution, 1975), 4, 6-7, 37, 45-46, 64-65, 75, 77-78, 80, 82-83, 85, 103-6, 116-18, 135-54, 314-15, 318-19. Any number of specialists in and outside the government struggle to influence foreign and defense policy; Morton H. Halperin, *Bureaucratic Politics and Foreign Policy* (Washington, D.C.: Brookings Institution, 1974), 17-19, 28-38. Similar divisions among experts may be found in virtually any governmental program area; Francis E. Rourke, *Bureaucracy, Politics, and Public Policy,* 2d ed. (Boston: Little, Brown, 1976), 119-20. My examples come from government organizations because I am familiar with them. The same reasoning probably applies to leaders in other fields, however; see, for example, Robert Aaron Gordon, *Business Leadership in the Large Corporation* (Berkeley: University of California Press, 1961), chap. 4; and Diane Borst and Patrick J. Montana, eds., *Managing Nonprofit Organizations* (New York: Amacon, 1977).

39. Simon, Smithburg, and Thompson, *Public Administration,* 108-10, 212-13, 388, 498-500, and chap. 14; Gordon, *Business Leadership in the Large Corporation,* pt. 2; Norton E. Long, "The Administrative Organization as a Political System," in *Concepts and Issues in Administrative Behavior,* ed. Sidney Mailick and Edward H. Van Ness (Englewood Cliffs, N.J.: Prentice-Hall, 1962), 110-21; Halperin, *Bureaucratic Politics and Foreign Policy,* chaps. 3-5.

40. Simon, Smithburg, and Thompson, *Public Administration,* 73-76; Harold J. Leavitt, *Managerial Psychology: An Introduction to Individuals, Pairs, and Groups in Organizations,* 2d ed. (Chicago: University of Chicago Press, 1964), chaps. 2-5; James G. March, "Bounded Rationality, Ambiguity, and the Engineering of Choice," *Bell Journal of Economics* 9, no. 2 (Autumn 1978): 587-608; and Jeffrey Pfeffer, Gerald R. Salancik, and Huseyin Leblebici,"Uncertainty and Social Influence in Organizational Decision-Making,"in Marshall W. Meyer and Associates, *Environments and Organizations* (San Francisco: Jossey-Bass, 1978), 306-32.

41. Only occasionally does this proposition appear flatly in the literature on organizations—e.g., in Victor A. Thompson, *Modern Organizations: A General Theory* (New York: Knopf, 1961), where hierarchy is severely criticized and proposals are advanced intended to replace "the grown-up kindergartens through which we now conduct our affairs" (197); in Robert Presthus, *The Organizational Society: An Analysis and a Theory* (New York: Knopf, 1962), where hunger for power and ruthlessness (chap. 6), empty posturing (173), and sheer physical size and strength (172) are said to carry weight in decisions; in Gordon Tullock,

The Politics of Bureaucracy (Washington, D.C.: Public Affairs Press, 1965), where bureaucratic leaders are portrayed as cynical, Machiavellian politicians; in Robert A. Caro, *The Power Broker: Robert Moses and the Fall of New York* (New York: Knopf, 1974), where Moses's success is attributed to intimidation, blackmail, misrepresentation, and disregard of ethics (12-19); and, more generally, in Jeffrey Pfeffer, *Power in Organizations* (Marshfield, Mass.: Pitman, 1981), where organizations are portrayed (esp. at 363-70) as arenas in which power politics is a major and growing element in decision making. See also, Alvin W. Gouldner, ed., *Studies in Leadership: Leadership and Democratic Action* (New York: Harper, 1950), esp. 3-66.

But my point is strongly implied by the large body of writing that describes organizational decision making as a process of accommodation among many kinds of participants employing all sorts of tactics; see Gordon, *Business Leadership in the Large Corporation,* 46-57, 75-79, 99-106; Dorwin Cartwright, "Influence, Leadership, Control," in *Handbook of Organizations,* ed. James G. March (Skokie, Ill.: Rand McNally, 1965), chap. 1; Neil W. Chamberlain, *Enterprise and Environment: The Firm in Time and Place* (New York: McGraw-Hill, 1968), 59-69; Halperin, *Bureaucratic Politics and Foreign Policy,* chap. 12; and Michael D. Cohen, James G. March, and Johan P. Olsen, "A Garbage Can Model of Organizational Choice," *Administrative Science Quarterly* 17 (1972): 1-25.

Such a process has benefits and advantages of various kinds; see page 51. Producing responses with the maximum probability of successfully meeting environmental challenges, however, would not seem self-evidently to be among its virtues.

42. Difficulties of implementation have received more attention from students of government than from students of other institutions; see James W. Fesler, *Public Administration: Theory and Practice* (Englewood Cliffs, N.J.: Prentice-Hall, 1980), chap. 8.

But these problems are not unique to government; that is why control is a familiar subject in treatises on management. See, for instance, Haynes and Massie, *Management: Analysis, Concepts, and Cases,* chap. 13, esp. the bibliography at 338-39; see also Philip H. Mirvis and David N. Berg, eds., *Failures in Organization Development and Change: Cases and Essays for Learning* (New York: Wiley, 1977). Manufacturing executives cannot take for granted that what happens on the production lines of their plants, in the dealerships on the sales front, and in the purchase of services and supplies will be wholly in accord with their wishes. Union chiefs cannot be certain the practices of their locals will match national policy in every respect. University presidents must take on faith a great deal of what goes on in the classrooms of their institutions. Leaders everywhere are, in a sense, hostage to their followers and subordinates.

Unquestionably the art of control has advanced rapidly, and deviations from top-level guidelines are kept within increasingly narrow limits, even in government (e.g., Kaufman, *The Forest Ranger,* pt. 2; and Stafford Beer, *Decision and Control: The Meaning of Operational Research and Management Cybernetics* [New York: Wiley, 1966]). Still they cannot be — or at least have not been — eliminated, if for no other reason than the tendency of subordinates to "overconform"; Scott, *Organizations,* 299-304. Consequently, even well-conceived adjustments to the shifting environment may not succeed.

43. Halperin, *Bureaucratic Politics and Foreign Policy,* chap. 13.

44. The costs of gathering, analyzing, and communicating information are a major impediment to the acquisition of data by decision makers. In addition, a host of other factors may block the flow of organizational information or distort its contents. Harold L. Wilensky, *Organizational Intelligence: Knowledge and Policy in Government and Industry* (New York: Basic Books, 1967).

45. Murray Edelman, *The Symbolic Uses of Politics* (Champaign: University of Illinois Press, 1964), chaps. 2, 3. See also his *Political Language: Words That Succeed and Policies That Fail* (New York: Academic Press, 1977), and Mirvis and Berg, *Failures in Organization Change and Development,* 1-3.

46. Simon, Smithburg, and Thompson, *Public Administration,* 42-43, 172, 439-41; and Downs, *Inside Bureaucracy,* 195-97.

47. See note 36.

48. The Roman Catholic church is a striking illustration, but other churches also have extended histories, and so do some states and their organs of government, cities, universities, charitable institutions, and even a few commercial, financial, and manufacturing enterprises.

But were there only a single extant organization tracing its origins back a long way I would be obliged to account for its long existence, in view of the hazards to existence presented by the roily environment. The one case would challenge my line of reasoning thus far.

CHAPTER 4

Q. Then How Do Long-Lived Organizations Attain Old Age?
A. They're Lucky.

The logic of the argument thus far leads me to the conclusion that the survival of some organizations for great lengths of time is largely a matter of luck. It seems to me such longevity comes about through the workings of chance.[1]

Why Chance Seems to Be the Answer

Speaking metaphorically, you might liken the environment to a reticular pattern of incessant waves constituting a perpetually varying net or screen sweeping continuously through the total aggregation of interlocked organizations that form in the human population.[2] The openings in the ever-changing screen constantly assume different shapes and sizes. At the same time, the organizations themselves are always changing as they try to avoid being swept away. If the two sets of changes are such that an organization can "fit" through the "holes" when the screen passes, the organization survives; if not, it is carried off. (Frequently the holes are so large that an organization will pass through no matter what it does or does not do; the efforts it makes in such instances are largely irrelevant, though it would be hard to convince the people making the efforts that this is the case. At times the mesh is so fine that nothing an organization does to save itself can succeed. At still other times, however, the actions of the organizations *are* the reason it is not swept away; it fits itself through the net. Accident may contribute as much as planning does to such happy outcomes; we understand very little about the screen, so our most ingenious strategies often lead to disaster instead of to good fortune.)

Let me stress again that this characterization of the relationship between organization and environment is purely metaphorical. It is not meant as a literal description of physical reality; no actual filter really sweeps through the world of organizations. The parallel is merely a literary device to suggest how environmental selection works, to make it easier to understand. It is a figure of speech that points up the apparently perpetual agitation of the surroundings in which organizations find themselves, and of the consequences of this condition for organizational survival. I would not bleed and die for this metaphor; I insist only that no portrayal, however vivid, can be accurate unless it captures the endless challenges to organizational existence and depicts the root causes of those challenges. And when it does, it makes survival and extinction matters of chance.

The result is a frequency distribution of organizational life spans, ranging from very brief existence to extremely long life. That is because some organizations would doubtless be carried off by the first environmental challenge they encounter; fewer would get through several waves of environmental pressure before succumbing; a smaller number would find themselves unscathed after many passes of the environmental filter; only a tiny handful would be fortunate enough to survive the countless sweeps of the metaphorical net to which they would be exposed in the course of extended life. The probability that the uninterrupted string of favorable outcomes necessary to produce an extremely long-lived organization willl occur must be reckoned as very low. But in a very large number of occasions (and the number of organizations forming, living, and dissolving in the world must be very great),[3] a rare low-probability event is not only possible; it should be expected. If, as I have suggested, there is no inherent reason why organizations should not continue indefinitely except for their extinguishment by the turbulence of their environment, one here and there could very well have the good fortune to escape every environmental hazard over protracted periods, and therefore to achieve extremely old age. If *many* did so, this explanation would not hold.

68

If an infinitesimal proportion of the whole population does, that is quite in accord with the laws of probability.

The leaders and members of surviving organizations are usually disposed to attribute the endurance of their organizations to their personal virtues and gifts rather than to the laws of chance. They are not guilty of hubris; they want their organizations to endure, they labor hard in that cause, they are rational, analytical creatures who can plan and calculate and learn, and so their belief that their efforts are responsible for their success is appealing. If I am right, however, we will find evidence inconsistent with this thesis when we gather and examine data of the following kind.

Why Skill Is Not the Answer

I anticipate, on grounds that I will get to shortly, that comparisons between organizations that survive and those that expire will in the vast majority of instances disclose no significant differences in their respective levels of ability, intelligence, or leadership talents. I don't *know* this to be the case; I am merely tracing out the logic of my argument. This proposition is advanced as part of my hypothesis, not as established fact. If the surmise turns out to be in accord with the facts, however, it will cast serious doubt on the belief that the cleverness of the people in long-lived organizations rather than the laws of probability account for the age of these structures.

The measures of ability, intelligence, and leadership in organizations must, of course, be independent of longevity itself. That is, we are not justified in concluding that surviving organizations are superior in these respects to those that die if we use survival as the evidence of superiority. That is circular reasoning. But if we can contrive independent indicia of the allegedly critical qualities, I predict that we will discover the failures and successes are very much alike in these regards. And if they are, if level of quality is not correlated with organizations' survival and extinction, the differences in their fates must be the result of something else. I think that "something else" will prove to be a probability function. This means that any individual organization's experience may be unique and puzzling, but as a member of a population of organizations, its mystery dissipates.

My grounds for expecting probability rather than skill to govern are twofold. One is that no two organizations, no matter how closely they resemble each other, are exactly alike. Nor are the conditions of their existence exactly alike. So the factors that impede adjustment to the volatile environment operate a little differently in each one, producing a range of responses to the same stimuli even when levels of ability are equal in the organizations affected. Indeed, even when organizations try to emulate each other, their endeavors come out differently.[4]

In the second place, much change in organizations is not consciously willed and may even be opposed by members and leaders.[5] Since organizations do not live in isolation but are involved in continuous commerce across their boundaries—commerce in people, ideas, information, and materiel—new concepts and outlooks invade them willy-nilly; they cannot be kept out. Moreover, new values, ways of thinking, and patterns of behavior also develop spontaneously inside organization boundaries as a result of specialization of skills and knowledge, shifts of power when individuals move within the formal and informal structures, and alterations of relationships when groups break up and assemble in new combinations and alliances. Consequently, organizations are likely to behave differently even if they are equally gifted and are confronted by the same problems.

Much of the time, I suppose, these differences between organizations may have little effect on their survival. At other times, however, even small differences may constitute the margin between survival and extinction when the environmental "network" of changes impinges on them. Thus, not only may organizations of comparable talent fare very differently from one another; logically, it would not be surprising if organizations blessed with outstanding gifts were sometimes extinguished while mediocre ones come through and flourish. If such things are not uncommon, as I postulate, chance must be the mechanism of selection.

What is more, chance could be the mechanism of selection even if talent plays a larger part than my speculations credit it with. For talent itself may be randomly distributed. Randomly does not mean evenly distributed, or that no organization is ever systematically and consistently more adept than most others at acquiring and developing talent. Rather, it implies that first-rate people can emerge from relatively obscure or undistinguished surroundings, rising to the occasion when dangers loom or opportunities beckon, furnishing an impetus or an innovation or the charisma that saves a threatened organization or lifts a hitherto mediocre one to great heights. By the same token, in a world where randomness reigns, great organizations may find themselves poorly led or badly mobilized at critical junctures and therefore unequal to the demands of the environment at a given moment, with fatal results. Thus, if I am right, the observer who bets on previously demonstrated quality of personnel to assure long organizational life or early death will often be surprised—more often than the observer who bets on sheer luck, good and bad.[6]

Perhaps the comparative data on organizational deaths and survival will not uphold this impression. If, however, survivors and succumbers are for the most part at much the same level of quality, claims for level of talent as the determinant of organizational endurance will have to give way to the statistics of probability.

Why Organizational Flexibility Is Not the Answer

Another explanation of notable organizational longevity is that organizational flexibility—the ability of organizations to change their structure or behavior or both, readily—permits the organizations endowed with it to cope with the dangers of the volatile environment and thus to survive indefinitely. They are the ones, according to this view of things, that are able to adapt easily whenever conditions warrant, to fit themselves through the environmental net, so to speak, regardless of the shape and size of its openings and the frequency of its passages. Highly flexible organizations could thus carry on for long periods, and, conversely, all long-lived organizations would turn out on close examination to be highly flexible.

Of course, flexibility itself could be the outcome of chance, a quality randomly distributed through the world population of organizations, the outcome of a host of fortuitous circumstances. The alleged advantages conferred on its possessors would then be nothing more than a fortunate break.

But it might also be portrayed as the product of organizational design. The leaders and members of any organization, aware of the great survival advantages of a generalized ability to respond to a host of environmental challenges, including unforeseen ones, would presumably strive to build this capability into their organization. Even if they were only partially successful, the likelihood of extended survival would seemingly be improved.

The trouble with flexibility as an explanation, regardless of its origins, is that it does *not* seem to me to assure long life for organizations. For one thing, flexibility is not costless; other organizational properties that also contribute to long life tend to diminish as it increases, so the *net* probabilities of long duration are not necessarily elevated when flexibility is maximized. For another thing, attributing organizational durability to flexibility implies that flexibility must remain at a fairly high level throughout the organization's existence (or else the organization would have been done in by the environment before it could attain old age), yet there is good reason to believe that flexibility *does* decline with age. Consider each point in turn.

THE COSTS OF FLEXIBILITY

As I see it, the costs of flexibility are of two kinds. One is its frequently perverse effect on the use of organizational resources. The other is its negative impact on the unifying power of internal organizational bonds.

Resource Costs. Flexibility implies the maintenance of the capacity to act in many different ways on short notice. To achieve this capacity, resources that might otherwise be fully employed in a particular immediately successful, rewarding fashion are partly withheld, or at least less than fully engaged, so that they can be shifted, restructured, and put to alternative and maybe previously unforeseeable uses. Once they are committed to a single option, switching to other options becomes extremely difficult.[7] The fear that such commitment may thus reduce adjustment capabilities is what leads to strategies meant to preserve flexibility, and the survival of many organizations possessed of flexibility is what keeps the fear of commitment alive in the population of organizations in the world.

Yet under certain conditions concentration of resources and wholehearted commitment to a single option may be advantageous. For example, specialty stores may be able to provide better service, higher quality products, and/or lower prices than general stores as long as the demand for their specialties holds up. (Of course, when fashions and tastes change rapidly, the general stores may be better off because they are equipped to satisfy a variety of demands rather than just one or a few.) Farmers who concentrate on one crop may get higher yields and superior returns while that product commands a good market; farmers who produce a variety of commodities in order to be prepared for shifts in markets may have difficulty competing. (But if shifts do occur, because supplies of the specialty crop go up or demand goes down, the diversified producers may be better able to withstand the new conditions.) Diversification increases flexibility, but it is not invariably an assurance of advantage. Sometimes the environment supports and rewards it; at other times, intense specialization, which implies less flexibility, may fare better.

Indeed, it is a paradox that maintaining flexibility can itself shut off options and impose limits on flexibility. An organization determined to maximize its fluidity locks itself into a circumscribed set of behaviors; it cannot do all sorts of things because they would be commitments and therefore would reduce its freedom of action.[8] It is not unconstrained; it merely subjects itself to self-imposed constraints. Since some of the potential uses of resources, foreclosed by emphasis on flexibility, may improve the chances of survival under some conditions, unyielding determination to preserve flexibility can be a cause of rigidity.

Costs in Disunity. Just as the preservation of flexibility sometimes leads to ineffectual employment of resources in response to environmental pressures, so also it can weaken the internal unity of organizations that is sometimes a key factor in triumphs over environmental adversity. Organizations are held together by forces that may poetically be compared to magnetism in the physical world.[9] That is, people contribute their time and energy, comply with directives and informal norms, and assist one another because of attractions to one another, to the collectivity, or both, and because they are repelled by possible alternative ways of life outside the collectivity. The bonds may be emotional, including love (of all by each, of a common leader, a common symbol, a common idea), hate, and fear; moral (the feeling that one *ought* to belong and obey and conform); expedient (self-interested calculation of the value of inducements offered by the organization as compared with the contributions asked of the individual and with inducements available elsewhere); habitual (behavior without emotional or moral roots, without any thought of alternatives, resulting from inertia and organizational indoctrination, training, and other forms of "brainwashing"); and physical (the walls of a prison, for instance). The mix and strength of the bonds vary from organization to organization, and even in a given organization over time. (In general, organizations united by a mixture of the first three seem likely to be more unified than those held together primarily by the last two, but perhaps a convergence of all the bonds produces the strongest union; these are empirical matters yet to be explored.)

Maximizing flexibility is likely to weaken many of the bonds, for the following reasons: It can weaken unity because it clashes with the cathexis strong bonds require. For example, the identification with a group or a symbol or an idea or a leader binding the members of an organization together may prevent shifts to new organizational patterns or practices or leaders when new conditions render the old features dysfunctional; hence, to preserve flexibility, an organization dare not let such attachments grow too strong. If the moral grounds of membership are stressed, changes of direction or practice for pragmatic reasons may be regarded as immoral and lead to wholesale desertions; to stay loose, organizations have to be cautious about invoking moral commitments. If expediential bonds are the main ones holding an organization together, they would be weakened by the obstacles to long-term contracts imposed by the demands of flexibility, even when such contracts are needed to keep key members from leaving. Similarly, the psychological conditioning that ties people to an organization is not easily undone when changed behavior is called for; organizations therefore do not always dare to ingrain habits as deeply as they might. So if an organization strives for flexibility, it frequently must eschew unifying devices that might otherwise be employed.

Admittedly, strong bonds are sometimes correlated *positively* with flexibility. For example, the organization members emotionally committed to a particular leader will follow him through occasional twists, turns, and even reversals of policy— through changes of structure, process, activity, product, and the formation or dissolution of alliances.[10] Similarly, people who join an organization mostly for the material rewards it offers will generally be relatively unconcerned about what the organization does as long as it continues to pay off.[11]

Even in such cases, however, flexibility is limited because leaders who make changes frequently are apt to exceed the tolerances of at least some of their followers, and the intensity of the commitment on the part of others will be diminished. Gradually there is an erosion of the bonds that permitted the freedom of action. A leader who keeps revising things may find his or her leadership challenged by a rebellious minority or weakened by increasing reluctance on the part of the majority to continue supporting the group at the top.[12]

Consequently, prudence constrains the most powerful leaders. Their ability to alter the structure or behavior of their organizations is an asset that can be used up. Thus they tend to draw on it sparingly. For them, too, flexibility can endanger organizational unity. For them, too, it is not a free good.

There may be times when the sacrifice of unity for flexibility is beneficial for an organization; I do not mean to suggest that the tradeoff always runs in one direction. But strong bonds could conceivably allow an organization to stay together through heavy weather that overwhelms organizations that opted for flexibility at the expense of cohesion; the power of nationalism, of family ties, of esprit de corps, and of romantic love, for example, is legendary.[13] Great cohesiveness can have great survival value; renouncing it for the sake of flexibility or anything else may cost an organization dear. Clearly, maximum flexibility is *not* an assurance of long life.

THE RAVAGES OF TIME

Nor is it plausible that organizational flexibility remains constant as organizations age (which the attribution of longevity to flexibility implies it would have had to do in order to keep old organizations viable for so long). The proposition has intuitive appeal if one assumes organizations accumulate experience and wisdom in the course of their long lives, thereby learning to dodge environmental bullets. But there are still stronger reasons for expecting organizations to grow more rigid, not more flexible, with time. The longer an organization has existed, the *smaller* its ability to change ought to be.

One reason for this correlation was mentioned in the preceding chapter: Over time, people develop vested interests in the status quo and therefore tend to resist change.[14] They exhibit proprietary attitudes toward the activities they perform, and oppose structural and operational modifications they perceive as reductions in the scope of their little empires. They become fearful of possible threats to the tangible and intangible rewards of their work. They get attached to, and comfortable with, what is familiar. They therefore do not shift readily. Time locks them into established modes of behavior.

In addition, precedents accumulate over time. Most organizations are not absolutely bound by precedent, but when they confront a problem they have a tendency to see how it or similar problems were handled in the past and to take those solutions as guides to current challenges. Folk wisdom, after all, is full of injunctions to stay on course: "Leave well enough alone." "If it is not necessary to change, it is necessary not to change." "If it ain't broke, don't fix it." True, aphorisms tend to come in mutually contradictory pairs; there are proverbs recommending innovation, too.[15] But as precedents pile up in organizations with the passage of the years, the forces of stability are likely to gain strength.

Indeed, organizations themselves generate such forces by training and indoctrinating their leaders and members to behave in organizationally prescribed ways.[16] Carried to the extreme, these endeavors would assure not only that their people conform to organizational commands and norms but that they are conditioned to act of their own volition, on their own impulses, in organizationally specified ways. Activities become routinized, automatic, ingrained. The more successful training and indoctrination are, the smoother, more reliable, more cooperative the organization is. At the same time, the more deeply implanted, the more thoroughly instilled, and the more completely programmed these behavior patterns are, the harder it is for the organization to change in response to environmental changes. For some reason, uprooting what has been learned often seems to be more difficult than implanting the original patterns.[17] So organizations often contribute to their own inflexibility.

Like individuals, organizations also amass sunk costs through past investments, future commitments based on past experience, and specialized knowledge and techniques. And since they contain organizations within their boundaries, and overlap other organizations, the veto points, accommodations, and negotiations that must be worked through grow more formidable with time.

That is why an organization's flexibility seems likely to decline as it gets older.[18] Along with the costs of flexibility, this tendency casts serious doubt on inherent flexibility as an explanation of the long life of old organizations.

Are Stable Ecological Niches the Answer?

Another possible explanation for the long persistence of a few organizations in the face of incessant environmental hazards is that pockets or pools of relative stability, like the backwaters

77

of turbulent rivers, may develop here and there in the environment and sustain unchanging organizations that form in, or migrate into, them. This explanation would be consistent with the observation that older organizations tend to be more inflexible than younger ones; rigidity would not be a liability in a stable setting. (Indeed, rigidity would be more advantageous than flexibility in a constant setting.)

This hypothesis presents serious difficulties, however. While it cannot be ruled out on logical grounds, invariance in a specific local environment in a world otherwise marked by continuous change is hard to imagine.

To be sure, such refuges might be chance phenomena, rare occurrences of exceedingly low probability among a great number of distinctive local environments; probability theory alone could be invoked to explain them. In that case, luck would be the chief factor in long organizational life, just as I have postulated. The stable pool hypothesis would not be identical with the version I have proposed because my version does not assume or require any islands of tranquility in the environment to explain longevity; the instability of the environment everywhere is there taken as a premise. Conceivably, then, the two explanations might ultimately diverge. It appears, however, that if they both rest primarily on chance, there are really no significant differences between them. Surviving for extremely long periods would be a matter of chance in either case.

On the other hand, they would be qualitatively different if stable niches are interpreted as functions essential to the very existence of every social system. Organizations performing such functions would naturally receive a large measure of support and protection from the other elements of the system and would therefore enjoy a high degree of security as long as the encompassing system as a whole continues, thus attaining a ripe old age. One could then argue that long organizational life is not a chance occurrence but a fixed requisite of the circumstances. The first organizations to acquire these responsibilities, or to have them thrust upon them, would not be easily displaced unless they experienced some exceptional internal failure. Hence, they could go on indefinitely.

A model of this sort would be determinate; long-lived organizations of specifiable traits would form inevitably, predictably. Perhaps when we know more about the organizational world, the existence of these inevitable, permanent niches will be obvious and undeniable. At the present stage of our knowledge, a probabilistic approach seems more fitting. The evidence that stable niches provide explanations of long-lived organizations often consists of the very existence of the organizations in question; the reasoning is frequently circular. If the circularity is to be avoided, independent methods of identifying stable organizational niches must be devised, and that is far from simple. Maybe long organizational life is *not* a product of chance, but if it is not, the specific causes have not been isolated. Unless and until they are, I shall construe it as a result of sheer good fortune.

Organizational Biology?

This interpretation of organizational experience bears certain resemblances to parts of the prevailing theory of organismic evolution. True, the approaches to the respective subjects come from different directions. Darwinian theory introduced the idea of species mutability into a body of thought that treated species as immutable; my contention is that organizations are frequently treated as highly plastic when their capacity for change is tightly constrained. But random variation and environmental selection figure prominently in both. One might therefore be tempted to infer that the same model is applicable to both. Are the mechanisms of biological and organizational evolution identical? Is the dynamics of the organizational world the same as the dynamics of the biological world? To these questions I turn next.

NOTES

1. See Armen A. Alchian, "Uncertainty, Evolution, and Economic Theory," *Journal of Political Economy* 58, no. 3 (June 1950): 211-21. The role

of chance in evolution, however, does not rule out the possibility of trends in the evolutionary process; see chapter 6.

2. The sieve metaphor did not find favor with one of the authorities on evolution, George Gaylord Simpson; see *The Meaning of Evolution: A Study of the History of Life and of Its Significance for Man* (New Haven: Yale University Press, 1949), 223. "It was a crude concept of natural selection," he wrote, "to think of it simply as something imposed on the species from the outside. It is not, as in the metaphor often used with reference to Darwinian selection, a sieve through which organisms are sifted. . . . It is rather a process intricately woven into the whole life of the group, equally present in the life and death of the individuals, in the associative relationships of the population, and in their extra-specific adaptations." I hope my characterization of the process of organizational selection, in which the properties of organizations themselves are important determinants of the environment and of adjustments to it, is not construed as a purely mechanistic concept of the kind Simpson deplores. I try to use it to suggest the striking variability and uncertainty of the organizational world, and the frustrations and disappointments associated with the efforts of organizations to cope with their situation. I do not mean to portray organizations and their environment as separate, independent forces, one active and the other passive. They are both part of the same fabric, the same intricate tapestry, which is why I resort to the network-of-waves metaphor. At any rate, it is only a point of departure for the discussion in the chapters that follow.

3. See page 45.

4. Richard R. Nelson and Sidney G. Winter, *An Evolutionary Theory of Economic Change* (Cambridge: Harvard University Press, 1982), 123-24, 267 68.

5. Herbert Kaufman, *The Limits of Organizational Change* (University, Ala.: University of Alabama Press, 1971), 41-44.

6. This position is akin to that described by Sidney Hook in his *The Hero in History: A Study in Limitation and Possibility* (Boston: Beacon Press, 1955), in which he says, at 14-15: "The role of the great man in history is not only a practical problem but one of the most fascinating theoretical questions of historical analysis. Ever since Carlyle, a century ago, proclaimed in his *Heroes and Hero-Worship* that, 'Universal History, the history of what man has accomplished in this world, is at bottom the History of the Great Men who have worked here,' the problem has intrigued historians, social theorists, and philosophers. . . . The Spencerians, the Hegelians, and the Marxists of every political persuasion—to mention only the most important schools of thought that have considered the problem—had a field day with Carlylean formulations. But in repudiating his extravagance, these critics substituted another doctrine which was just as extravagant although stated in lan-

guage more prosaic and dull. Great men were interpreted as colorful nodes and points on the curve of social evolution to which no tangents could be drawn. What is more significant, they overlooked a possible position which was not merely an intermediate one between two over-simplified contraries, but which sought to apply one of Darwin's key concepts to the problem; namely, *variation*. According to this view, the great men were thrown up by 'chance' in the processes of natural variation while the social environment served as a selective agency in providing them with the opportunities to get their work done." (Copyright 1943, by Sidney Hook. Reprinted in 1950 by Humanities Press by special arrangement with Sidney Hook. Excerpt reprinted by permission of Humanities Press Inc., Atlantic Highlands, N.J. 07716.) See also Kenneth E. Boulding, *Ecodynamics: A New Theory of Societal Evolution* (Beverly Hills, Calif.: Sage, 1978), 218-19, 238-40.

7. Herbert A. Simon, *Administrative Behavior: A Study of Decision-Making Processes in Administrative Organization* (New York: Macmillan, 1947), 66, 95-96, 120; Herbert A. Simon, Donald W. Smithburg, and Victor A. Thompson, *Public Administration* (New York: Knopf, 1950), 427-28; and Kaufman, *The Limits of Organizational Change,* 29-30.

8. Kaufman, *The Limits of Organizational Change,* 71.

9. Ibid., 116-17. See also E. Wight Bakke, *Bonds of Organization: An Appraisal of Corporate Human Relations* (New York: Harper, 1950), p. 8, chap. 7, and Appendix C for a different approach.

10. For example, many members of the Communist Party of the United States, whose ideology held that the policies of the Soviet Union were good for working people throughout the world, followed Soviet leadership through switches from attacks on liberal reformers to efforts to make common cause with them; then, during the period of the Hitler-Stalin nonaggression treaty, back to attacks on liberal supporters of the Allies in Europe; followed by alliance with them when the Soviet Union was invaded by Germany; and once again back to hostility when the cold war began. See Earl Latham, "Communist Party, United States of America," in *Dictionary of American History,* rev. ed. (New York: Scribner's, 1976), 3:143-46, esp. 144-45.

By the same token, nationalist considerations have permitted millions of citizens to follow their governments through alliances with recent enemies and hostilities with recent allies.

11. The things about which they don't care were said by Chester I. Barnard to fall in their "zone of indifference"; see *The Functions of the Executive* (Cambridge: Harvard University Press, 1938), 168-69. Economists have developed formal models of indifference curves and indifference maps; Paul A. Samuelson, *Economics,* 11th ed. (New York: McGraw-Hill, 1980), 416-23.

12. Eventually, the frequent changes in policy by the Communist Party of the United States alienated most of its erstwhile members; see Latham,

"Communist Party, United States of America." And more than two millennia earlier, even the mighty Alexander the Great, whose troops worshiped him, had to abandon his conquest of India when the weary, homesick soldiers simply declined to go on; see R. Ernest Dupuy and Trevor N. Dupuy, *The Encyclopedia of Military History from 3500 B.C. to the Present* (New York: Harper & Row, 1970), 52. That is why Herbert A. Simon, emphasizing the power of subordinates, preferred to call their willingness to obey orders their "zone of acceptance" rather than their "zone of indifference" as Barnard did (see note 11); *Administrative Behavior,* 12.

13. Perhaps this is one of the reasons that recent empires—which were presumably more diversified than individual nations—were disrupted by economic, social, and political changes that nations managed to survive, and why the immensely versatile armies of powerful nations have often had difficulties in dealing with guerrilla forces specialized to the local environment and motivated by intensely held, shared beliefs.

14. Gerald Zaltman and Robert Duncan, *Strategies for Planned Change* (New York: Wiley, 1977), chap. 3. See also chapter 3, note 36, in this volume.

15. "Nothing ventured, nothing gained." "Opportunity knocks but once." "Build a better mousetrap, and the world will beat a path to your door." "Strike while the iron is hot." "He who hesitates is lost."

16. A number of students of human behavior have remarked on this phenomenon; see Robert K. Merton, *Social Theory and Social Structure,* rev. and enlarged ed. (New York: Free Press, 1957), 197-98. For a recent illustration, see Herbert Kaufman, *The Administrative Behavior of Federal Bureau Chiefs* (Washington, D.C.: Brookings Institution, 1981), 118-22 and, more generally, 91-124.

17. The difficulty is especially acute when "the new situation presents stimuli that are similar or identical to those in a previous learning situation but demands dissimilar or opposite responses . . . ; the previous response persists and retards acquisition of the new one.

"Furthermore, the old responses are likely to reappear even after the new ones have been mastered and under the worst possible circumstances. Thus, in times of crisis or under stress, people may abandon present or new modes of coping with problems and revert to earlier patterns." Bernard Berelson and Gary A. Steiner, *Human Behavior: An Inventory of Scientific Findings* (New York: Harcourt, Brace and World, 1964), 161.

18. This position reverses my suggestion in an earlier work that age and flexibility in organizations are probably correlated positively rather than inversely; see Kaufman, *The Limits of Organizational Change,* 98-99. At the time, I could think of no other explanation for the long life expectancy of older organizations as compared with younger ones (see chapter 2, note 4, in this volume), in a volatile environment. But the proposition that flexibility grows with age appears to be contravened

by evidence, logic, and authority. Therefore, I abandon the initial position in favor of the one presented in this chapter. (Note, however, the recent finding in Thomas W. Casstevens, "Population Dynamics of Governmental Bureaus," *The UMAP Journal* 5 [1984]: 194, that age and survival are unrelated. This finding implies that age and flexibility are unrelated or, if they *are* related, that flexibility and survival are unrelated. The question remains open.)

CHAPTER 5

Q. Does This Argument Equate Organizational Evolution with Biological Evolution?

A. No. Organizations Are Not Organisms.

No form of life on earth stands outside the process of biological evolution. Even species that have lived in ecological niches for tens of thousands of years, changing very little in all that time, came to their present status via natural selection, and survive there because of evolutionary dynamics. Every living thing is part of it.

If organizations were living things, it would stand to reason they too would be products of, and governed by, the dynamics of biological evolution. Indeed, organization theory may well be a branch of biology; after all, aren't human organizations constellations and communities of organisms? And don't organisms and organizations resemble each other in striking ways?

But while students of organizations may well have more to learn from biological metaphors than from analogies to other disciplines, they would not be justified in equating organizations and organisms. Searching for literal equivalencies can be misleading. In the last analysis, although organizations are a way of mobilizing the energies of living things (as well as other forms of energy), they are not organisms. The characteristics of organisms should therefore not be imputed to them.

Why Organisms and Organizations Are Sometimes Equated

Treating organized collectivities of people as though they were equivalent to organisms "writ large" is a longstanding tradition.[1] At times, the analogy is explicit, comparing organizational structures and operations to nerves, brains, sensory organs, memory, circulatory systems, and other bodily systems.[2] Tacit or metaphorical comparisons, however, are probably more common. All sorts of sentiments are attributed to collectivities — aspiration, fear, hope, hate, anger, pleasure, and others.[3] Organizations are often portrayed as goal-seeking, reasoning, and calculating. They are treated as consciously and rationally adaptive, planning their actions after surveying and evaluating the options open to them. Some students speak of them, usually heuristically, as though they were not only organisms but human beings of an extraordinarily intelligent kind.[4]

Indeed, one could even identify certain likenesses in organizational and organismic self-replication. Many organisms reproduce asexually, dividing into parts, each of which is or becomes a complete new individual; the number of parts may be two or more and they may be equal or unequal in size. Major illustrations are fission (mitosis), fragmentation (pieces that break off a parent and grow into whole organisms), sporulation (the formation of spores that can grow into organisms), and budding (protuberances that grow on cells until a wall appears and separates the protuberances from the originals, at which point the protuberances turn into new cells).[5] Could not the departure of a person or a group of persons from an organization to establish an independent copy of the original be likened to some of these procedures? Are not the founding of colonies by citizens of a city-state or a country, the creation of subsidiaries by companies, the establishment of new units of government on the model of existing units, and similar operations all types of organizational reproduction?

The analogy might even be extended to sexual reproduction by a determined comparatist.[6] In sexual reproduction, half the information (instructions) controlling the characteristics of an organism are transmitted from each parent to the offspring. Don't organizations also receive a blend of instructions and programs influencing their structure and behavior? This information is transmitted in a variety of ways — observation of the structure and behavior of other organizations; systematic sharing of information through both technical and general publications; movement of personnel from organization to organization; and an individual's simultaneous memberships in overlapping organizations. (Sexual reproduction is especially beneficial in a changing environment because it introduces variations into the gene pool, increasing the chances that at least *some* offspring will survive under the new conditions. Asexual reproduction is advantageous in stable environmental niches, where things stay virtually the same for protracted intervals, and where preservation of genetic instructions without variation maintains a good match between organism and surrounding.)[7] One could argue that the combination and recombination of instructions transmitted to organizations are tantamount to the reshuffling of genetic information in the gene pools of organisms.

Carrying this logic to its extreme, the next step would be to describe selection of organizations for survival or extinction as identical to the natural selection and evolution of organisms. But there are such differences between organisms and organizations that the often-useful metaphor can do as much harm as good.

Why They Are Not Equatable[8]

For one thing, the living structures of which organizations are composed are orders of magnitude apart from those that make up organisms. Most of the components of organisms are so tightly bound to the entities of which they are part that each dies very quickly if it is separated from the organism or if the organism as a whole dies. Nor can they detach themselves from one organism and take up life in another (although massive surgical intervention can now move certain organs or tissues from one person to another). By and large, the cells, tissues, and organs of a specific organism are linked exclusively to that organism. The degree of interdependence between the parts and the whole is very high.[9]

But the members of organizations, individually and in groups, readily survive separation from entities of which they are part and even survive their death, though the pain of the sundering may be severe. *Total* isolation, it is true, can be shattering; [10] fortunately, it is rare. People, as individuals or in clusters, usually can move easily from one organization to another, frequently on their own initiative and also when separation is imposed on them. Bonding in organizations is much looser and weaker than in organisms.[11] This difference is bound to affect the respective histories of the two kinds of structures.

Another significant difference is the discreteness of organisms as compared with organizations. By and large, the physical components of an organism are contained exclusively within that organism. But organizations share components because each person, and even whole groups, can belong to more than one organization at a time. Multiple memberships create overlapping and nesting organizations. Organizations intersect and are interlocked in ways that have no counterparts in the organismic world. It would be astonishing, therefore, if many of the laws governing their structure and behavior were not different from those applicable to organisms, despite some notable similarities between the two.

Moreover, there are no precise organizational counterparts to organismic generations. We can hardly speak of "parents" and "offspring" in the same sense in both cases. "Descent" has a different meaning in each of these contexts.

Finally, the precision and reliability with which genetic instructions are transmitted from parents to offspring in organismic reproduction are many times greater than the communication of organizational information when organizations resembling other organizations form by any means of replication. The cells in living things all contain the same genetic material; although different segments of the material are activated to produce different kinds of tissues and functions, the effects are dependable. Against this, the components of organizations — individual persons and groups of people — are not uniformly programmed, instructions are often progressively modified as they pass along communication channels, and communications are not uniformly interpreted by those who receive them. Organisms produce such remarkably true copies of themselves that species were long believed to be immutable. The best copies of organizations, on the other hand, are only rough approximations of the prototypes on which they are fashioned.

The conclusion implied by these premises is that similarities between organismic parents and their offspring result from transmitted structural and behavioral programming to a far larger extent than do similarities between organizations that succeed each other. That is, while transmitted programs and environment together shape both organizations and organisms, organisms retain their programs over generations while the retentive mechanisms of organizations are not as strong or sure or enduring. Consequently, the immediate environment of any given organization accounts for a larger proportion of its characteristics than would be the case with any given organism.[12] That is why it would be misleading to assume that the process of biological evolution has an exact counterpart in the organizational world. What happens in the organizational world, if my logic to this point is valid, must be different.

A Different Analogue:
The Organizational Medium

That is why I was drawn to an analogy between organizations and the structures thought by many scientists to have preceded and paved the way for the appearance of life on earth.[13] This analogy strikes me as more appropriate than the likening of organizations to organisms because, although the putative structures antedating life, like organizations, did not have the precise, dependable mechanism of heredity that marks living things, they too formed and disintegrated and were replaced by other structures in profusion. According to this theory, a process of "chemical evolution" went on despite the absence of true genetic inheritance. Advances in the durability of the structures were never completely lost because chemical changes producing the improvements were discharged into the medium around the structures when they dissolved and were then incorporated into newly forming structures. This method of transmission was uncertain and highly imperfect, but a good many biologists and biochemists believe it worked well enough to sustain a progression toward more complex (and ultimately living) structures. This model seems to me to fit the characteristics and experience of organizations quite well.

More specifically, the picture drawn by the biologists and biochemists who hold this view is as follows: Three billion years ago, the Earth was warm and wet; the seas were rich in atoms of carbon, hydrogen, oxygen, and phosphorous; and the surface was bombarded with energy in the form of lightning and sunlight unfiltered by ozone (because the ultraviolet-light-filtering ozone layer in the atmosphere did not develop until plant life built up the volume of oxygen in the air). The application of energy to the elements under these conditions produced compounds in the warm seas and forced the compounds into various combinations. Over millions of years the compounds became more complex and abundant until the seas were teeming with molecules of all kinds. And still chemical reactions continued, driven by the energy pouring down, until the substances aggregated into droplets of varying composition and chemical reactions depending on the constituents they chanced to contain.[14] Most of the droplets probably broke up soon after forming; the life spans of many of them may have been measured in minutes, if not seconds. But their chemical activity and the release of their contents back into the water presumably changed the chemical composition of their liquid environment gradually over millennia. As new droplets formed, some among them would have to have been more complex than their predecessors because they were composed of more complex material. Some would therefore prove more stable and durable, renewing themselves for longer periods by developing faster reaction times as a result of their changing makeup, and growing larger as they incorporated new materials into their own structures. The drops were ultimately competing with each other for the nutrients around them. Since the speed of chemical reactions in drops of different composition varied, and since the growth rates of drops varied as a consequence, say some theorists, some must have fared well, others poorly. The more dynamic ones—those with the fastest chemical reaction speeds and the highest growth rates—must have become an ever-larger proportion of all such drops in the seas. The chemistry of the seas and the chemistry of the drops thus would have spiraled upward in complexity as each changed the composition of the other. Eventually, in this view, droplets formed with the capacity to reproduce themselves, and the character of the planet was dramatically altered. Because these new forms could multiply so rapidly, they came to occupy every nook and cranny of the globe, displacing the more primitive chemical droplets.[15]

Recent experimentation points to another possible line of chemical evolution. According to this theory,[16] the predecessor of the first cells capable of reproduction was not a coacervate droplet but a progression from micromolecular to macromolecular to supramolecular systems that were assembled into protocells.[17] Preprotein was present on the primitive Earth, and only contact with water would have been necessary to set the process in motion. While the transition from the protocell to the contemporary cell has not been fully explained, experiments suggest developments that could have bridged the gap.[18] Presumably, with each advance, more materials contributing to the appearance of the next stage became available in the water. Once organisms appeared, they quickly consumed the matter that had the capacity to undergo conversion to organisms, so the spontaneous appearance of life did not repeat itself; from then on, evolution took place through living forms.[19] But the primordial process, say the adherents to this theory, probably occurred many times in many places, in parallel, and the products always resembled each other because they were governed by the same chemical properties and were formed of similar raw materials.[20] The evolution from simplicity to complexity began before the cell and continued afterward. Thus, though this model differs from the coacervate-drop model in major respects, they both find the origins and history of life in a medium that not only provides the initial conditions required to set it in motion but itself changes in ways that make further development possible.

The whole prevailing doctrine of chemical evolution has recently been challenged by experts who argue that no kind of evolution is possible until a long-term hereditary mechanism begins to work.[21] They think that a method of transmitting information developed long before the appearance of the first "naked gene," and that a primitive organism thus equipped had to be in existence to shelter and sustain the first genetic (RNA-like) material to appear fortuitously; otherwise, the material would not have survived. That is, there can be no genotype without an accompanying phenotype, and in this view a pre-existing phenotype had emerged which the first genetic material could take over. The chemistry of this process is quite different from that proposed for chemical evolution.[22]

I am not qualified to take sides in these debates; fortunately, I don't have to. I am not claiming a one-to-one correspondence between what occurred on the surface of the Earth in the distant past and what is happening in the development of organizations in our own time; I am merely pointing out where the idea I am here setting forth came from. Even if the concept of evolution without genetics should eventually prove inapplicable in many respects to the chemical history preceding the emergence of life, it seems to me to describe and illuminate remarkably well the workings of the world of human organizations. I am not saying the evolution of organizations will inevitably produce a transfiguration of organizational forms comparable to the presumed emergence of organisms from inorganic matter; I shall deal later with the organizational implications of the analogy, which is richly suggestive. My point here is that the parallels are striking.

Like the postulated primordial molecules and droplets, organizations form out of the medium in which they will exist, from which they cannot be separated, and on which they draw constantly. (More generally, I mean by the "medium" of a dynamic structure the set of ingredients surrounding the structure from which it obtains the materials and energy and behavioral programs incorporated into its substance when it forms and from which it continues to obtain them as long as it endures. The medium for organizations consists of people, elements of their culture, and energy, the building blocks of which the organizations themselves are composed.)[23] Whatever forces bring the components of any organization together, boundaries develop and activities that sustain the entity commence. It is born and it begins functioning in ways that keep it alive.

That is why I described organizations as engines that, once started, tend to keep running indefinitely unless something stops them.[24] The tendency to keep going manifests itself in intensified activity of one sort or another when the operation of the engines is impeded. In organisms, such heightened, sometimes frantic, activity is often construed as evidence of an "instinct of self-preservation," but it seems to occur in organizations too and may also have appeared in primordial droplets. Bear in mind, however, the limits on the degree to which organisms and organizations can (and probably the primordial droplets could) adjust to changing conditions. Most organisms are bound by their genetic heritage and can therefore adapt only in the course of successive generations. The molecules and droplets were governed by the laws of chemical reactions and could continue only within a narrow range of conditions. Each organization, as we have seen, is constrained by a host of factors.[25] In a dynamic environment all three kinds of structures would suffer heavy casualties.

Droplets created experimentally are the most fragile of the formations, breaking up and re-forming readily.[26] Microparticles generated experimentally by students of molecular evolution are more stable, but also are vulnerable to environmental variation.[27] Organizations come apart easily and return their constituent elements to the medium, where the elements—people, knowledge, machines, and so on—are reabsorbed into new organizations. Organisms are different; although they too die in large numbers and are recycled, their germ plasm has great durability and survives almost intact long after particular structures in which it is imbedded disintegrate.[28] That sets them apart. Organizations, although made up of living organisms, seem to me more like the hypothetical primordial precursors to the first cells than like organisms. Their structural and behavioral patterns are transmitted indirectly, largely through the enveloping medium, to the organizations that succeed them, with the result that much accumulated information is lost or altered or adulterated, and alien patterns intrude when the successors form. This process of transmission is loose, jumbled, and messy compared to organismic inheritance. Well, who ever said everything in this world has to be tidy?

Just as some primordial formations chanced to be constituted differently than others or to be surrounded by a more hospitable environment than others, or both, and therefore survived longer, so do chance factors determine which organizations will get through the environmental screen and how long they will last. Indeed, the primeval structures could be said to be involved in competition with others of their kind for the elements needed to sustain them; in parallel fashion, organizations, because they form profusely, may possibly press on the available supplies of personnel, material, and energy in the environment and thus may be in competition with all other organizations, even those whose activities are quite different from theirs.

Since organizations, organisms, and the chemical predecessors of life thus are all governed by their own laws, separate courses of development should be anticipated. On the other hand, since the mechanisms of development in all three cases are evolutionary, it is reasonable to think there might be many common features. And so we are brought to the next question about the differences and similarities in the different units of evolution: How are their evolutionary mechanisms alike and how are they dissimilar?

NOTES

1. Francis W. Coker, *Organismic Theories of the State: Nineteenth Century Interpretations of the State as Organism or as Person* (New York: Columbia University Press, 1910); Robert Redfield, ed., *Levels of Integration in Biological and Social Systems* (Tempe, Ariz.: Jaques Cattell Press, 1942); John R. Kimberly, "The Life Cycle Analogy and the Study of Organizations: Introduction," in John R. Kimberly and Robert H. Miles et al., *The Organizational Life Cycle: Issues in the Creation, Transformation, and Decline of Organizations* (San Francisco: Jossey-Bass, 1980), 6-13; and Michael Keeley, "Organizational Analogy: A Compar-

ison of Organismic and Social Contract Models," *Administrative Science Quarterly* 25, no. 2 (June 1980): 337-62.

2. See the works cited and criticized by Robert K. Merton in his *Social Theory and Social Structure,* rev. ed. (New York: Free Press, 1957), 48, note 51. See also Karl W. Deutsch, *The Nerves of Government: Models of Political Communication and Control* (New York: Free Press, 1966), esp. 81.

3. Keeley, "Organizational Analogy," 337. See also, Bertram M. Gross on organizational "personality" in his *The Managing of Organizations: The Administrative Struggle* (New York: Free Press, 1964), 2:487-90.

4. Julian Feldman and Herschel E. Kanter, "Organizational Decision Making," in *Handbook of Organizations,* ed. James G. March (Skokie, Ill.: Rand McNally, 1965), 614-49. See also Donald W. Taylor, "Decision Making and Problem Solving," ibid., 48-86; Bo Hedberg, "How Organizations Learn and Unlearn," in *Handbook of Organizational Design,* vol 1, *Adapting Organizations to Their Environments,* ed. Paul C. Nystrom and William H. Starbuck (New York: Oxford University Press, 1981), 3-27.

5. "Reproduction," *The New Columbia Encyclopedia* (New York: Columbia University Press, 1975), 2304-5.

6. For example, Roger D. Masters, "Genes, Language, and Evolution," *Semiotica* 2 (1970): 295-320.

7. See note 5.

8. For a much fuller discussion of the similarities and differences between social and biological evolution, see Edgar S. Dunn, Jr., *Social and Economic Development: A Process of Social Learning* (Baltimore: Johns Hopkins University Press, 1971), 80-112; and Donald T. Campbell, "On the Conflicts Between Biological and Social Evolution and Between Psychology and Moral Tradition," *American Psychologist* 30, no. 12 (December 1975): 1103-26.

9. See Redfield, *Levels of Integration in Biological and Social Systems;* Merton, *Social Theory and Social Structure,* 25-28; and Donald T. Campbell, "Common Fate, Similarity, and Other Indices of the Status of Aggregates of Persons as Social Entities," *Behavioral Science* 3, no. 1 (January 1958): 14-25.

10. Bernard Berelson and Gary A. Steiner, *Human Behavior: An Inventory of Scientific Findings* (New York: Harcourt, Brace and World, 1964), 252.

11. Organization theorists refer to the comparative autonomy of individuals, groups, and suborganizations in organizations as "loose coupling"; Howard E. Aldrich, *Organizations and Environments* (Englewood Cliffs, N.J.: Prentice-Hall, 1979), 76-86, 325-27; W. Richard Scott, *Organizations: Rational, Natural, and Open Systems* (Englewood Cliffs, N.J.: Prentice-Hall, 1981), 107-8, 254-58.

12. That is not to say organisms are governed exclusively, or even mostly,

by their heredity. The argument is only that *compared to organizations,* organisms' retentive mechanisms are especially powerful. The effects of environment may nevertheless be significant in the growth and behavior of any organism.

13. For example, J. H. Rush, *The Dawn of Life* (Garden City, N.Y.: Hanover House, 1957), esp. chap. 6.

14. A. I. Oparin, *The Origin of Life on the Earth,* 3d ed. (New York: Academic Press, 1957), 353-63. Other chemical theories have been advanced —for example, J. W. S. Springle, "The Origin of Life," in Society for Experimental Biology, *Symposia* 7 (1953), on *Evolution,* 1-21; Sidney W. Fox and Klaus Dose, *Molecular Evolution and the Origins of Life,* rev. ed. (New York: Marcel Dekker, 1977); A. G. Cairns-Smith, *Genetic Takeover and the Mineral Origins of Life* (New York: Cambridge University Press, 1982).

15. Oparin, *The Origin of Life on the Earth,* 363; Rush, *The Dawn of Life,* 101-2; and Mahlon B. Hoagland, *The Roots of Life: A Layman's Guide to Genes, Evolution, and the Ways of Cells* (Boston: Houghton Mifflin, 1979), 39-40.

16. Fox and Dose, *Molecular Evolution,* 262-65.

17. Ibid., 242, 256, 262.

18. Ibid., 252-54, 264.

19. Ibid., 11.

20. Ibid., 255.

21. Cairns-Smith, *Genetic Takeover,* 10-21, 45-60, 60-77, 261-64, 300-302; John Noble Wilford, "New Finding Backs Idea That Life Started in Clay Rather Than Sea," *New York Times,* 3 April 1985.

22. Ibid., 79-83, 121-24.

23. For a fuller description, see page 144.

24. See pages 17-19.

25. See pages 76-77.

26. Fox and Dose, *Molecular Evolution,* 226.

27. Ibid., 208.

28. "Samuel Butler's famous aphorism, that the chicken is only an egg's way of making another egg, has been modernized: the organism is only DNA's way of making more DNA"; Edward O. Wilson, *Sociobiology: The New Synthesis* (Cambridge: Harvard University Press, 1975), 3. See also Richard Dawkins, *The Selfish Gene* (New York: Oxford University Press, 1976), 25: "A body is the genes' way of preserving the genes unaltered." (The arguments proceeding from these figures of speech have been criticized for departing from the Darwinian concept that individual organisms, not genes or groups of organisms, are the units of biological evolution; Stephen Jay Gould, *The Panda's Thumb: More Reflections in Natural History* [New York: Norton, 1982], 85-92. But the passages quoted do highlight vividly the extraordinary continuity of genetic inheritance, a mechanism of transmission obviously quite

different from that which operates in organizations. I present them only to invoke authority for this specific point, not to enter an arena of contention among professional biologists.)

CHAPTER 6

Q. Then Do They Evolve Differently?
A. Yes, but Not in Different Directions.

Any processes in which chance plays such a large part would seem unlikely to proceed consistently in one direction. Rather, they would be expected to meander chaotically as a result of changing course randomly.

And indeed there has been a lot of meandering in the history of both organisms and organizations. Forms flourished and then disappeared. Distinguishing attributes and clusters of attributes set forms apart from one another, but then some of the differences declined. Elaborate and complicated forms established themselves and dominated their ecological zones, only to be displaced by less intricate forms. Evolutionary history is a record of fits and starts, twists and turns, dead branches and new growth, and multiple lines of development.[1] But it is the record of a trend, too.

The Trend toward Complexity

Organismic and organizational evolution (and even postulated preorganic chemical evolution) have apparently produced forms encompassing within their boundaries discrete structures and substructures, each engaged in different activities that collectively contribute to the continuation of the forms. Many of the most intricately internally differentiated forms have been of much more recent origin than the earliest forms of life. Hence, if such internal differentiation of structure and function, which is frequently attended by great interdependence of the parts (sometimes so great that the failure of any one disrupts or terminates the operations of all the others), is adopted as a definition of complexity, then we may say that evolution is marked by a general tendency toward greater complexity.[2] Chemically, in the primordial seas, the trend is said to have showed itself in the developments that ended in the emergence of life.[3] Biologically, it is exhibited in the progression from unicellular to multicellular to social forms of life.[4] Organizationally, it manifests itself in the progression from the most primitive postulated forms of human group to the intensively specialized, actively interrelated, large-scale associations of more recent times.[5]

I do not mean that more recent forms, biological or organizational, are always more complex than earlier ones. Such an inference would be patently false. Many ancient species were apparently extremely complex,[6] and so were many ancient civilizations.[7] Nor do I mean that more complex forms invariably displace or dominate less complex ones; different levels of complexity exist side by side, each dominant in a different sphere.[8] And more complex forms are not necessarily better adapted or more adaptive than simpler ones.[9] Some are, some are not.

All I am saying is that, in the course of very long stretches of time, forms of life and of organization more complex than anything seen before will be added to the mix of forms inhabiting the globe. In the short run this trend may not be apparent; indeed, the reverse may sometimes seem to obtain. But if you consider the whole span of life and all of what we know and surmise about organizations during human history and prehistory, the trend line can be discerned. It seems to be an inherent thrust of the evolutionary process. To be sure, if some cataclysm should destroy all of the more complex forms at once, the old heights of complexity might never be reached again. But the *thrust* toward greater complexity would reassert itself because of the "built-in" dynamics of the evolutionary process— and, if no further cataclysms occurred, forms of greater complexity than even the previous high points *might* eventually be added to the mix.[10]

If such a consistent thrust obtains, it introduces yet another enigma into the argument. Earlier, I attributed organizational evolution to the operation of chance in organizational variation and environmental selection. But if it is a matter of chance, how does it happen to proceed in a particular direction? How can randomness produce a distinct trend line?

The Explanation for Living Things

The reconciliation of randomness and the trend toward complexity among living things rests on the notion that living things occupy zones or spheres in the environment that are filled up as a result of their capacity to reproduce, so that the zones will not sustain any more of them.[11] At the same time, the shuffling of genes in the gene pool and mutations of the genes from other causes produce individuals showing some variance with the modal traits of the population. Historically, some of these variations resulted in higher rates of survival among offspring possessing them, and the traits gradually diffused through the population of the species.[12] Some of the members then were able to enter zones into which they could not have penetrated before. There they frequently found themselves in competition with earlier occupants, sometimes to their detriment, sometimes to their advantage. Occasionally they found themselves in virtually unoccupied zones. Usually such pioneering meant that the survival rate of their unchallenged offspring was very high. The fortunate species with the lucky traits therefore came eventually to fill the hitherto uninhabited zones.[13]

Not all unoccupied ecological zones necessarily favored greater complexity.[14] A good many apparently did, however; it took fairly complicated creatures to make some of the transitions. As newly occupied zones filled up, the process was repeated. Indeed, the expansion of life and the emergence of new forms kept creating new environmental spheres into which newer forms could shift. (I should add that many environmental spheres may occur in the same geographic area; the spheres are sets of conditions, not separate territories. Making a transition to a new zone therefore did not always mean a change in locale.) Presumably, then, simplification as well as increases in complexity also took place; some zones probably were supportive of *simplified* variations of existing species. But at least in some instances, greater complexity was advantageous, and life forms of higher complexity were therefore added to the mix from time to time. In this fashion, chance and environmental selection combined to yield the tendency toward complexity, a thrust intrinsic to biological evolution according to current concepts.

The Explanation for Organizations

The same tendency is evident in the dynamics of organizations and in hypothesizing about the primordial chemical substances preceding life (which I consider analogous),[15] but the *way* it comes about for both these types of structures is much less complicated than the one just described for organisms because of the absence of a genetic mechanism. Yet it moves in the same direction, toward greater complexity.

For organizations, and presumably for preliving chemical structures, the increase in complexity results from changes in the respective media out of which they form. We have already seen how the chemistry of the early Earth was likely to have engendered increasingly complex substances, which in turn made for more chemical complexity, whereupon new substances formed out of the more complex medium and nurtured still more elaborate structures and processes. If this hypothesis is right, the general thrust toward greater complexity would have been unremitting and self-reinforcing, though some specific, limited, temporary interruptions may have occurred.

The same logic applies to the tendencies of organizational evolution. The medium of which organizations are formed and in which they grow, defined earlier as consisting of people, culture, and energy, is changed by the activities of organizations. Probably few organizations have profound effects, and no single one by itself, no matter how influential, is likely to have more than a marginal impact on the medium. In time, as new organizations replace old ones, and as outputs of new organizations pass through the environment into existing ones, the accumulation of changes mounts to substantial proportions. Gradually all components of the organizational medium are altered.

People, for instance, are modified by the organizational roles into which they are cast, the training to which they are subjected, the sociology of the groups to which they are exposed.[16] If the organizations of which they are members break up, or if they leave for other reasons, they bring with them at least some of the outlooks, values, expectations, aspirations, habits, and knowledge acquired in the original setting. The next organization into which they are drawn must inevitably be affected by their behavior and modes of thinking when they come in.[17] This does not *invariably* add to the complexity of the receiving organization. In general, however, as more and more people in a society move through numbers of organizations, they are likely to expect and demand more elaborate and sophisticated services, benefits, protections, standards, and treatment than before, and thus to push up the level of organizational complexity. A taste of complexity, or even knowledge of it, often whets the appetite for it, so that once it is experienced it tends to keep things from going back to where they were.[18] And even if the trend levels off or reverses in one part of the worldwide organizational medium, it usually continues upward in other parts, eventually impinging on the quiescent sectors and starting it upward in them once again. Progress is not uniform or steady, but the average in the organizational world is probably always rising.[19]

Frequently technology alone explains the rise in complexity.[20] Once the railroad, the telephone, the automobile, radio and television, the airplane, and the computer were introduced, for example, they brought in their wake hosts of new organizations, practices, services, products, and life styles. These developments quickly strained the original technology to its limits and provided stimuli to improvements. The improvements in turn encouraged more people to use the inventions. The increased usage, in its turn, was a stimulus to further improvements. Once set in motion, the process tended to drive itself.

The increase in complexity follows also from the greater varieties of skills and knowledge and resources needed for even small operations to get started and sustain themselves in such a setting. Coordination and the conduct of relations with other highly specialized organizations, necessitated by finer and finer division of labor and rising interdependence in the system, become separate functions. To be sure, simple organizations also continue to form and survive. But as the setting changes, organizations of greater sophistication, encompassing specialized tasks and units that have to be kept in balance with one another, make their appearance. New organizations of even modest size come to exhibit these characteristics. Complexity pervades the environment.

Gradually, what goes on even at some distance begins to have a noticeable impact on every organization and individual in a complex culture. Predictably, organizations (mostly governmental in the United States) specializing in the prevention and/or relief of hardship resulting from such impacts spring up.[21] They then constitute another factor in the environment with which other organizations must cope as they conduct their own activities; dealing with this aspect of the environment often gives rise to service units specializing in this set of tasks only. Thus complexity breeds complexity within a culture.[22]

Even the production and consumption of energy within and by organizations contributes to increasing complexity. As organizations deplete the readily available sources by using up stored deposits, exhausting the soil, and polluting water and air, new specialties within old organizations and new organizations performing new functions are generated by the need to acquire new supplies, restore old ones, and reduce the amounts of energy consumed for each unit of goods or services produced.[23] Although most people and organizations simply go on as they did before, depending on supplies furnished through the new developments, complexity in the system is raised by the appearance of the new structures, activities, and forms of dependence. Ultimately the effects spread everywhere.

In summary, no matter which aspect of the organizational medium is examined, the result is always the same. The medium keeps growing more and more complex as a consequence of all these factors. As new organizations form and grow in it, some of *them* will be more complex and carry on more complicated activities than their predecessors—partly as a deliberate strategy for survival in the changing environment, but largely because the people and things and knowledge they incorporate into themselves change and complicate them whether they know or want it. These modifications engender more complexity in the medium, setting off a new round of development. Even organizations that die contribute to it as the release of their complex contents into the medium "thickens" it and affects their successors. Thus, while complex organizations do not necessarily grow at the *expense* of simpler ones, ever-more-complex forms keep appearing and joining those already in existence.

Where Is the Trend Heading?

To say that increasingly complex organizations will join the mix of organizations in the course of time is not to say we can predict accurately the structure and activities of the additions to the organizational population. The character and dimensions of the new organizations depend on so many simultaneous factors that often they can be discerned only after they have emerged. Knowing that there is a trend toward organizational complexity in evolutionary processes does not enable us at this stage of our knowledge to forecast what form the new complexity will take.

Yet one has the feeling that extrapolating the trend line must yield *some* idea of the state toward which the organizational evolutionary process is tending despite our inability to say exactly what will happen next. *Is* the process heading toward an end we can dimly determine? That is the question to which we now turn.

NOTES

1. With respect to organismic evolution, see George Gaylord Simpson, *The Meaning of Evolution: A Study of the History of Life and of Its Significance for Man* (New Haven: Yale University Press, 1949), chap. 15, esp. p. 243: "Whatever criterion you choose to adopt, you are sure to find that by it the history of life provides examples not only of progress but also of retrogression or degeneration." For an instance of the

same phenomenon in one kind of organization, see R. M. MacIver, *The Web of Government* (New York: Macmillan, 1947), 162-74, esp. 163, where, speaking of forms of government, the author declares: "In times of great crisis a sort of political 'Gresham's law' comes into operation—the more primitive form drives out the more complex."

2. This premise was one of Herbert Spencer's central theses. See Robert L. Carneiro, "Herbert Spencer," *International Encyclopedia of the Social Sciences* (New York: Macmillan and Free Press, 1968), 15:122-23; and Richard Hofstadter, *Social Darwinism in American Thought*, rev. ed. (Boston: Beacon Press, 1955), 37. But it has also had more recent exponents: see notes 3, 4, and 5; the discussion of "general evolution" in Marshall D. Sahlins and Elman R. Service, eds., *Evolution and Culture* (Ann Arbor: University of Michigan Press, 1960), chap. 2; A. G. Cairns-Smith, *Genetic Takeover and the Mineral Origins of Life* (New York: Cambridge University Press, 1982), 89-95; and Peter A. Corning, *The Synergism Hypothesis: A Theory of Progressive Evolution* (New York: McGraw-Hill, 1983). See also Herbert A. Simon, "The Architecture of Complexity," *Proceedings of the American Philosophical Society* 106, no. 6 (December 1962): 470-73.

3. J. H. Rush, *The Dawn of Life* (Garden City, N.Y.: Hanover House, 1957), 140-42, particularly the comment, 141-42, that "we may . . . conceive of a chain of evolution that began when . . . atoms were formed (or earlier!) and has continued unbroken, from complexity to greater complexity, down to the intricate higher organisms of the present day." See also A. I. Oparin, *The Origin of Life on Earth*, 3d ed. (New York: Academic Press, 1957), 92-102, 351-63; and Sidney W. Fox and Klaus Dose, *Molecular Evolution and the Origin of Life*, rev. ed. (New York: Marcel Dekker, 1977), 264.

4. Robert W. Redfield, ed., *Levels of Integration in Biological and Social Systems* (Tempe, Ariz.: Jaques Cattell Press, 1942); Caryl P. Haskins, *Of Societies and Men* (London: George Allen and Unwin, 1952), chap. 2, "The Tide to Complexity." But see notes 1 and 14 in this chapter; the progression is not necessarily smooth.

5. Robert E. Park, "Biological and Social Systems," in Redfield, *Levels of Integration in Biological and Social Systems*, 217-40, esp. 223-26; Haskins, *Of Societies and Men*, chap. 12; Edgar S. Dunn, Jr., *Economic and Social Development: A Process of Social Learning* (Baltimore: Johns Hopkins University Press, 1971), 91-94; and David Warsh, *The Idea of Economic Complexity* (New York: Viking Press, 1984).

6. "Whether Recent man is to be considered more complex or more independent than a Cambrian trilobite will be found subject to qualifications and dependent on definitions"; Simpson, *The Meaning of Evolution*, 117. "It would be a brave anatomist who would attempt to prove that Recent man is more complicated than a Devonian ostracoderm"; ibid, 253. "What modern biochemistry, molecular biology, and cell biology

have taught us is that life even at the level of bacteria is an astonishing phenomenon, of enormous complexity; the biochemical pathways in bacteria involve hundreds of enzymes, each coded to carry out its special catalytic function, all these are integrated into a highly coordinated system capable of drawing in metabolites from the most unpromising surroundings and of completely replicating themselves in a matter of minutes rather than hours"; E. J. Ambrose, *The Nature and Origin of the Biological World* (Chichester [West Sussex], England: Ellis Horwood, 1982), 164.

7. A good many societies before the Christian era attained high levels of complexity and sophistication; see, for example, Herbert Heaton, *Economic History of Europe,* rev. ed. (New York: Harper, 1948), chaps. 2, 3; Ralph Linton, *The Tree of Culture* (New York: Knopf, 1955), pts. 5 through 10; and C. D. Darlington, *The Evolution of Man and Society* (New York: Simon and Schuster, 1969), pts. 2, 3, 4, and 8.

8. "We do not . . . find successive dominance between, say, Osteichthyes, Aves, and Mammalia. All three are dominant at the same time, during the Cenozoic and down to now. Taking the animal kingdom as a whole, it is clearly necessary to add insects, molluscs, and also the 'lowly' Protozoa as groups now dominant. If one group had to be picked as most dominant now, it would have to be the insects, but the fact is that all these groups are fully dominant, each in a different sphere." Simpson, *The Meaning of Evolution,* 245-46

   Similarly, cultures characterized by many different patterns coexist in the modern world; they are the data of cultural and social anthropology. The question of dominance, to be sure, is quite ambiguous as the impact of industrial culture reaches every corner of the globe. Moreover, cultural relativists challenge the notion that nonliterate societies are "simpler" than literate societies; Melville V. Herskovits, *Man and his Works: The Science of Cultural Anthropology* (New York: Knopf, 1948), 70-75. Still, as I have defined complexity, it appears that degrees of complexity vary from one contemporaneous culture to another, and that many of the less complex ones do indeed hold their own in some respects despite the overwhelming material superiority of industrial civilization, at least for the time being—if for no other reason than that industrial powers have not sought to submerge them, but also perhaps because the benefits of the effort would not be worth the costs.

9. More complex units are not necessarily more wondrous than simpler ones. Many of the things that go on in complex ones have direct counterparts in simpler units; they are just performed in different ways. Many of the relationships between simpler units and their environment are as finely balanced and as strong as those established by complex units. There is as much to marvel at in simpler forms as in their complex neighbors. "Above the level of the virus, . . . the simplest living unit

III

is almost incredibly complex"; Simpson, *The Meaning of Evolution,* 15.

As regards adaptiveness: "Every organism is adapted to some particular way of life. . . . [T]here is no way to decide that one adaptive type is more adaptive than another"; ibid., 249. Indeed, some "simpler" forms, such as viruses, have proved capable of adapting rapidly to changed conditions, while the fossil record abounds with cases of complex creatures that could not.

10. "If man were wiped out, it is extremely improbable that anything very similar to him would ever again evolve, although I cannot see that even this is altogether impossible. The exact ancestral forms are gone. The whole intricate sequence of biological and physical conditions that gave rise to man certainly will not be repeated with very close approximation. Yet it remains true that manlike intelligence and individual adaptability have high selective value in evolution and that other animals have a conceivable basis for similar development." Simpson, *The Meaning of Evolution,* 327.

11. Ibid., 114-15. See also Cairns-Smith, *Genetic Takeover,* 89-91. For a more general discussion of the role of chance and determinism in evolution, see the remarks by Ernst Mayr (26), and Garland E. Allen (365-66), in *The Evolutionary Synthesis: Perspectives on the Unification of Biology,* ed. Ernst Mayr and William B. Provine, (Cambridge: Harvard University Press, 1980); and Ernst Mayr, *The Growth of Biological Thought: Diversity, Evolution, and Inheritance* (Cambridge: Harvard University Press, 1982), 519-20.

12. Theodosius Dobzhansky, *Mankind Evolving: The Evolution of the Human Species* (New Haven: Yale University Press, 1962), 141-42.

13. But the process seemed to contain a built-in ratchet preventing it from reversing: "In the filling of the earth with life, some spaces were filled first, filled well and adequately, leaving neither reason nor possibility for refilling by animals of later development." Simpson, *The Meaning of Evolution,* 114.

14. "Progressive simplification, as well as progressive complication, may accompany progress." Simpson, *The Meaning of Evolution,* 254.

15. See pages 91-96.

16. See page 77 and notes 16 and 17 in chapter 4.

17. Herbert Kaufman, *The Limits of Organizational Change* (University, Ala.: University of Alabama Press, 1971), 41-44.

18. "Dynamic forms of adaptation consist of acceptance of change, social learning, striving for further change, and a continued stepping up of levels of aspiration as each higher goal is achieved"; George Katona, Burkhard Strumpel, and Ernest Zahn, *Aspirations and Affluence: Comparative Studies in the United States and Western Europe* (New York: McGraw-Hill, 1971), 187. Or, as Sam M. Lewis, Joe Young, and Walter Donaldson put it in song in 1919 after the doughboys of World War I

returned from France, "How Ya Gonna Keep 'em Down on the Farm After They've Seen Paree?"

19. This proposition is conjectural, of course. Conceivably, the appearance of highly complex forms might be offset by the appearance of new simple ones, or by increases in the numbers of existing simple forms, holding average complexity constant or even lowering it. Indeed, complexity might even generate countervailing simplicity in the organizational world by reducing task complexity and skill requirements through routinization. My impression, however, is that differentiation of structure and function and the degree of interdependence of parts in the worldwide human population tend to rise on the average, even though the level of particular individual human tasks is frequently reduced to mindless monotony for a time. If we can devise appropriate measures, we will be able to choose between these conflicting possibilities. An approach to such measurement is proposed in the Postscript, pages 144-48.

20. Kenneth E. Boulding, *The Organizational Revolution: A Study in the Ethics of Economic Organization* (New York: Harper, 1953), 25-26. See also note 11 in chapter 3.

21. Herbert Kaufman, *Red Tape: Its Origins, Uses, and Abuses* (Washington: Brookings Institution, 1977), chap. 2.

22. See also note 26 in chapter 3.

23. The Department of Energy, the Environmental Protection Agency, the Forest Service, the Soil Conservation Service, the Water and Power Resources Service, and the Fish and Wildlife Service are among the agencies of the federal government established to deal with energy and resource problems. In addition, private firms formed to produce equipment devised to comply with the requirements of some of these agencies and the laws they administer, to inform and advise private individuals and groups on their obligations under the new laws and regulations, and to represent them in proceedings before some agencies. Interest groups mobilized to support and oppose the administrative bodies. International institutions formed to coordinate actions and policies crossing national boundaries. State and local governments set up counterpart agencies. Within existing organizations, units specializing in these problems and relationships appeared. That is, scarcity engenders organizational complexity no less than abundance does.

CHAPTER 7

Q. Where Will the Organizational
Trend End?

A. It Probably Won't.

Organizational complexity as it is defined in the preceding chapter seems to be correlated with organizational size and scope. That is, organizations of greater complexity appear to contain more people and suborganizations and activities and perhaps area within their boundaries than less complex ones do. While it is not extraordinary for suborganizations to be more complex than the organizations that contain them, size and broad scope will more often accompany organizational complexity.[1]

Two hypotheses—which are complementary rather than contradictory and linked rather than independent—though each stresses a different set of factors, could explain such a correlation. One is the proposition that organizations tend to grow as large as organizational skill and capacity enable them to grow.[2] The division of labor and specialization of function—which is to say, the complexity—characteristic of technologically advanced organizations permit them to grow very large, do many different things, and elaborate their structures.

The other hypothesis, suggested earlier,[3] is that organizations are averse to uncertainty and generally respond to it with a built-in tendency to expand their boundaries or centralize, depending on whether the source of uncertainty is outside or inside their boundaries. Since there are always new sources of uncertainty outside the borders, no matter how large they grow, they always tend to expand. Since every expansion incorporates new sources of uncertainty within their boundaries, they always tend to centralize. The combination of expansion and centralization makes for organizations of great complexity.

If organizational complexity, size, and scope are indeed associated and increasing, they would seem to lead unavoidably to some predictable final point. It is hard to avoid the impression that extrapolating the trend must tell us how it will end.

In fact, different people foresee or imagine different outcomes. The end point, if there is one, is not all that clear. The trend may encounter limiting forces that check it or change its character.

Possible Steady States

If the organizational world generates forces that run counter to present tendencies, those forces would presumably govern until *they* in turn set off another set of forces counter to themselves, which would drive things in the original direction until the counterforces were once again activated. Thus, the entire system would oscillate within strict limits.

One possible form of such a steady state was imagined by George Orwell in *Nineteen Eighty-Four*.[4] He saw the tendency toward complexity, size, and centralization resulting in three vast superstates into which most of the world would be drawn. None of the superstates by itself could overcome and absorb or suppress the others. Any two, however, could conquer the third. But both of the surviving allies would then live in fear of each other. So just at the point where the third superstate is about to be defeated, one of the imminently victorious pair breaks up the alliance and joins with the intended victim against the former partner. The new alliance holds until the former partner is on the verge of elimination, whereupon one of the new pair switches sides. All three would be locked for all eternity in a pas de trois from which they could never advance, never retreat, and never escape.

Another form of steady state would be continuous cycling, with larger, more inclusive organizations appearing and growing, then breaking apart and starting to grow again from the original level at which the cycle began. Large organizations might break up for a variety of reasons. Their size and complexity and centralization might reach a level at which their organs of central direction and coordination are overburdened and collapse. Suborganizations, driven by the same impulses as the more inclusive organizations of which they are part, may secede in numbers from the system, de facto if not de jure. Specialization and interdependence among the suborganizations may become so intense that any small subset of the membership can immobilize the entire system. The encompassing organization, under any of these conditions, would gradually lose the capacity to perform the activities that hold the system together and maintain its boundaries. The components would thereupon be released back into the medium, and the overall organization would vanish. (That is, changes in the internal environment of organizations as well as in the world outside their boundaries can lead to their demise.) But when containing organizations come apart, the process that led to their appearance in the first place would generate pressure toward new containing organizations—not identical with the originals, for conditions would have changed in the interim, but organizations tending toward more and more complexity, inclusiveness, and centralization. They in their turn would grow until the inescapable centrifugal forces they harbor tear them open, whereupon regeneration would recommence. Thus, an observer with some perspective on the process would see entities form, grow, break up, form, grow, break up, repeating their cycles ad infinitum, endlessly cycling between a lower limit of utter atomism and an upper limit of near-organic unity.

Like the Orwellian stalemate, this situation is a steady state. But it oscillates more. Whereas the Orwellian system always looks the same, with participants merely exchanging roles, the repetitive cycle changes appearance at different stages. Neither model is static. Both are continuously in motion—within their respective established bounds, it is true, but always in motion.

Possible Changes of State

There is no a priori reason why such steady states cannot appear anywhere in nature. Linear trends, however, seem more commonly to lead to changes of state in which a different dynamics takes over. In the physical world, for instance, raising the temperature changes solids to liquids and liquids to gases; reducing the temperature of gases, which diminishes the volume of the gases by amounts that would result in their total disappearance at absolute zero, turns the gases to liquids and even to solids long before absolute zero is reached. In the biological world the increasing complexity of the droplets in the primordial seas set the scene for the appearance of life, which proceeded along an evolutionary path through a different mechanism. One may conjecture, therefore, that the process of organizational evolution could end in a comparable transformation.

Some scholars,[5] extrapolating from the forces that pressed life along the path from unicellular to multicellular organisms and from multicellular organisms to social groupings, seemed to suggest (though not unanimously or without qualifications) that social structures are evolving toward a worldwide organization so highly integrated, coordinated, and cohesive that it could be thought of as almost equivalent to a single organism. Instead of getting stuck in a steady state short of this ultimate form, the process, they speculated, would eventuate in a comprehensive, strongly unified entity.

If such a threshold were crossed it would surely constitute a momentous change of state for organizations, if not for humans as organisms. It would probably reduce the unpredictability of the environment for suborganizations within the encompassing system, doubtless reducing the death rate and thereby altering profoundly the pattern of evolution through organizational births and deaths that obtained earlier. It would also diminish the diversity of suborganizations within the system, which could mean less likelihood that any part of the system would survive sudden, drastic changes in the worldwide biological or physical surroundings. It would contract the self-containment of subsystems so that failures anywhere in the intricate web of relationships could wreak havoc everywhere.

The internal environment would probably be stabler and less stressful for people and suborganizations. A highly coordinated system might also be capable of achievements beyond the capacity of a fragmented, turbulent organizational world. Yet it might restrict the freedom of individual people by subjecting them to massive controls on mind and behavior designed to ensure their compatibility with the requirements of such a tight-knit organization;[6] spontaneous, nonconforming action, and independent thinking that leads to such action, would be a threat to the system's order and tranquility if not to its very existence (unless originality, creativity, and unorthodoxy were carefully confined and manipulated, as in Aldous Huxley's *Brave New World*).[7]

So the dynamics of an all-encompassing and organically integrated organization would probably be of a different order from the dynamics of the organizational world described earlier. If it were to come about, it would not do so suddenly, in an instantaneous transformation. Rather, it would occur gradually, perhaps even imperceptibly. But differences in degree can accumulate into differences in kind, as the wavelengths of light shade by indistinguishable increments from red to yellow. One might never be able to say, "This is the point at which the new dynamics took hold." Still, comparing the core attributes of the old process with the core attributes of the new would leave no doubt that a significant change had indeed occurred.

The same would be true of another possible change of state to which the process of organizational evolution might conceivably lead—domination of the evolution of human institutions and organizations by the autonomous evolution of intelligent machines. Commentators have discussed this possibility for generations, and von Neumann many years ago described how machines might be programmed not only to make copies of themselves but to introduce random changes into their design and assembly so that variation and "natural" selection could continue to provide for adjustments to conditions for which planned preparations could not be made because they could not be foreseen by the most sophisticated intelligence, human or artificial.[8] If the evolutionary process hypothesized here works as suggested, such an outcome is a reasonable, if not a highly probable, prospect.

For a long time machines have constituted a major, and probably a steadily increasing, component of the medium in which organizations form. Much of the rising complexity of the medium can be ascribed to the number and sophistication of machines within it. To be sure, they are artifacts of the human mind and servants of the human will. But each new machine capability engenders organizational activities that soon reach the capacity of existing technology. Then, with the aid of the existing machines, new ones of greater capacity are created. So the capabilities keep advancing, caterpillar fashion, stretching to the limit, then bringing up the base and stretching forward again. And as organizations continuously form in the medium of human culture they incorporate these advanced capabilities into their structure and activities.

As a result, machines play a larger and larger part in the life of every human organization. Many organizations depend so heavily on the machines that mechanical breakdowns virtually paralyze them. As machines take over more and more computational chores the dependence grows. If they can be designed to think, or merely to approximate human thought, however crudely, the dependence will become greater. If they can be designed to learn from experience and thus to modify their original programs or to formulate new programs for their own operations, more and more organizational activities will fall within their province. If they can produce new designs for new machines, they will begin to evolve on their own. The proportion of machines in the medium will rise and new organizations will be networks of machines operated and supervised by other machines, with people standing by for emergencies.[9]

That is not to say people will become completely superfluous or obsolete; eventually that could conceivably be their fate, but not necessarily. Organizations may remain a mix of humans and machines well into the remote future. The process of *organizational evolution,* however, could become largely a function of machine development. While organizations would presumably continue to form and re-form in their changing medium, succumbing to the environment or surviving much as they have always done, the variations among them would be determined mostly by differences in, or introduced by, their machines.

Maybe organizational evolution in which machines occupy such a central position would still be governed by the dynamics that governed it in the past. But it could, on the other hand, take on a completely different complexion. After all, organizational self-replication would certainly have different attributes, the character of exchanges with the environment and even within organizational boundaries would not be the same, and the bonds holding organizations together would acquire a different quality. With so many properties of the evolving units changing it would not be surprising if the properties of the process of evolution itself also underwent fundamental revision, a transformation of state rather than just a modification of an existing state.

Even if such a transformation occurs, this transition, like the appearance of the predicted worldwide organization, would probably not be noticed by many people until the new process had been in effect for a long while. It is apt to come gradually (compared to human generations, not to geological spans) rather than suddenly, and therefore to be visible only when the distance from the past is greater. Recognized or not, however, it could mark the beginning of a line of development whose unfolding we can hardly now imagine.

The appearance of an all-inclusive organization and the shift to evolution through machines might each be regarded as an end to the evolutionary process postulated in these pages and the commencement of something to be plotted on a different set of coordinates. But neither of them would constitute a final chapter in the history of organizations. In that sense they would not be endings but only stages in a story that goes on as long as people and organizations themselves exist.

Possibilities Undreamed Of

This survey of conceivable outcomes of organizational evolution that come quickly to mind as one grapples with the last question in this catechism probably does not exhaust the possibilities. The process is capable of so many twists and turns and surprises that it may head in directions totally unforeseen by those who study it. Its course is not fixed.

Of one thing, however, we may be virtually certain. It will not eventuate in a *static* state. The organizational world, even in a steady state, will be full of commotion. It is so fluid and agitated, in fact, that the prospect of its remaining fixed for long periods within unchanging limits is improbable. Rather, driven by the forces with which we are already familiar and by others we have yet to discern, and steered by chance interactions between organizations and their environment, it seems likely to continue to evolve as far into the future as the most visionary among us dare to speculate about.[10] It is an unending challenge to humankind's imagination, reason, and persistence.

NOTES

1. "Organizational size is positively associated with structural differentiation. Studies of a wide variety of organizations show reasonably consistent and positive associations between size of organization and various measures of structural differentiation. . . . Larger organizations tend to be structurally more complex"; W. Richard Scott, *Organizations: Rational, Natural, and Open Systems* (Englewood Cliffs, N.J.: Prentice-Hall, 1981), 237. The studies to which the passage refers are cited in it. More generally, see 235-44 for a review of the literature and a discussion of ambiguities and deficiencies of the data and the findings.

As regards my assumption that suborganizations can be more complex than the organizations that contain them, I rely on the prevalence of associations of quite complicated organizations of many kinds. Some are associations of governments, such as the United Nations, the Council of State Governments, and the National League of Cities. Some are groups of firms, such as the Association of American Railroads, the Air Transport Association of America, the American Bankers Association, the American Sugarbeet Growers Association, the Associated General Contractors of America, and the Association of the Wall and Ceiling Industries. Some are constellations of colleges and universities, like the American Council on Education, the Association of American Universities, the Association of Independent Colleges and Schools, the Association of Schools and Colleges, and the National Association of State

Universities and Land-Grant Colleges. Some are clusters of nonprofit agencies, including the National Organization of Non-Profit Homes, the American Association of Museums, and the American Library Association. The great labor unions are federations of internationals and nationals and locals of occupation- or industry-specific workers. Even individual departments of government are often holding companies of specialized bureaus; see Herbert A. Simon, Donald W. Smithburg, and Victor A. Thompson, *Public Administration* (New York: Knopf, 1950), 269-72. The examples are legion. Not *all* the constituents of these clusters are more complex than the associations containing them. But *many* of the constituents doubtless encompass a greater variety of activities and differentiated structures than do the separate and distinct organs of the containing organizations—or so I postulate. This is another indicated line of research.

2. Kenneth E. Boulding, *The Organizational Revolution: A Study in the Ethics of Economic Organization* (New York: Harper, 1953), chap. 2.

3. See pages 42-44. See also Herbert Kaufman, *The Limits of Organizational Change* (University, Ala.: University of Alabama Press, 1971), 79-86.

4. New York: Harcourt, Brace, 1949.

5. Robert Redfield, ed., *Levels of Integration in Biological and Social Systems* (Tempe, Ariz.: Jaques Cattell Press, 1942); and Cesare Marchetti, "From the Primeval Soup to World Government: An Essay on Comparative Evolution," *Technological Forecasting and Social Change: An International Journal* 11, no. 1 (1977): 1-7. See also Roderick Seidenberg, *Posthistoric Man: An Inquiry* (Chapel Hill: University of North Carolina Press, 1950).

6. "[T]he dominant minority will create a uniform, all-enveloping, superplanetary structure, designed for automatic operation. Instead of functioning actively as an autonomous personality, man will become a passive, purposeless, machine-controlled animal. . .'"; Lewis Mumford, *The Myth of the Machine: Technics and Human Development* (New York: Harcourt, Brace and World, 1967), 3. In the symposium reported in Redfield, *Levels of Integration in Biological and Social Systems,* however, there were differences of opinion on this score; see "Introduction," 23-25.

7. Garden City, N.Y.: Doubleday, 1932. See also Seidenberg, *Posthistoric Man,* chap. 5, esp. pp. 176-79.

8. Described by John G. Kemeny, "Man Viewed as a Machine," 132-46 of a collection of essays assembled by the editors of *Scientific American* and published in 1955 by Simon and Schuster under the title, *Automatic Control.* "Could such machines go through an evolutionary process?" asks the author at 145-46. "One might design the tails [containing their programs] in such a way that in every cycle a small number of random changes occurred (e.g., changing an 'on' to an 'off' in the code or vice versa). These would be like mutations; if the machine could still produce offspring, it would pass the changes on." See also, W. Ross Ashby,

"The Self-Reproducing System," in *Aspects of the Theory of Artificial Intelligence: The Proceedings of the First International Symposium on Biosimulation, Locarno, 1960,* ed. C. A. Muses (New York: Plenum Press, 1962), 9-18; and Robert Jastrow, *The Enchanted Loom: Mind in the Universe* (Austin, Tex.: S & S Press, 1981), 158-68.

9. See Harold Borko, ed., *Computer Applications in the Social Sciences* (Englewood Cliffs, N.J.: Prentice-Hall, 1962), esp. chap. 2, "Do Computers Think?"; Edward A. Feigenbaum and Julian Feldman, eds., *Computers and Thought* (New York: McGraw-HIll, 1963); John Diebold, ed., *The World of the Computer* (New York: Random House, 1973); Avron Barr and Edward Feigenbaum, eds., *The Handbook of Artificial Intelligence* (Los Altos, Calif.: William Kaufmann, 1981), 1, 3-10, 14-17; Igor Aleksander and Piers Burnett, *Reinventing Man: The Robot Becomes Reality* (New York: Holt, Rinehart and Winston, 1983); and Robert U. Ayres and Steven M. Miller, *Robotics: Applications and Social Implications* (Cambridge: Ballinger, 1983), esp. chap. 8.

The rate of computer development is suggested by the comment of the editor of *The World of the Computer,* John Diebold, that "the computer is entering its second quarter-century and its fourth generation" (3), and by a story by Michael Schrage, "U.S. Spurs Computer Supremacy," in the *Washington Post* of 1 April 1983, describing the race between Japan and America "to create a so-called 'Fifth Generation' computer" (or "supercomputer") by 1990. See also Edward A. Feigenbaum and Pamela McCorduck, *The Fifth Generation: Artificial Intelligence and Japan's Computer Challenge to the World* (Reading, Mass.: Addison-Wesley, 1983).

For forecasts of robot population growth in the United States, see H. Allan Hunt and Timothy L. Hunt, *Human Resource Implications of Robotics* (Kalamazoo, Mich.: W. E. Upjohn Institute for Employment Research, 1983), chap. 2.

10. See Loren Eiseley, *The Unexpected Universe* (New York: Harcourt, Brace, and World, 1969), esp. 36-37, 45-47, 83, 179.

Q. Of What Use Is the Hypothesis?

A. It May Help Us Get Our Bearings.

The point of departure for this exercise was the question, Why do organizations die? If the hypothesis does nothing else, it provides a plausible and, as I explain in the Postscript that follows this Epilogue, a testable answer. If it accomplishes that much, I will feel amply rewarded for my effort.

Certainly the hypothesis is not much help to leaders and members of organizations seeking solutions to immediate problems confronting them. It does not tell them what they can do to improve their positions, prolong the lives of their organizations, escape from their difficulties. It does not yield a set of strategies and tactics that will promptly reduce the hazards they face. It is not a blueprint for triumph over adversaries or for increasing efficiency. It is not a formula for security and contentment.

It does not tell people what they ought to do day by day. So what good is it?

If it is right, it can be as useful as an accurate map. A map does not tell people where to go. But it lets them know some of the obstacles standing between where they are and where they are headed. Such knowledge does not ensure the success of their journey; it merely permits them to avoid some hardships they might otherwise suffer and perhaps reduce the odds of fruitless endeavor.

In similar fashion, a valid hypothesis puts in proper perspective constellations of factors with which we must come to grips as we pursue our ends, prepares us for what we are likely to encounter, and alerts us to courses of action that will probably prove unproductive. Some theories have done more than this; they have also made people aware of potentialities that they could not previously have imagined and suggested solutions to problems that would otherwise have continued to baffle humankind. But these are rare. Theories need not be so revolutionary to be useful. They can serve us even if they simply let us know how parts of the world work and thereby rescue us from labors that would otherwise come to naught.

The utility of the hypothesis advanced here, in its present state, is of this limited kind. Because its assumptions and its postulated dynamics are a little different from most others in the field, it may illuminate some dark corners that the others, however brilliant their light, leave in shadow. Another candle in the darkness can't hurt and might help.

In particular, the hypothesis calls attention to the positive side of organizational death. Many people believe that some organizations continue beyond the time when their existence can be justified, and this conviction has even led to efforts to increase the death rate of governmental agencies in a number of jurisdictions.[1] The advocates of this view, however, apparently mean that only organizations they consider unsatisfactory should expire, not that substantial turnover throughout the organizational world is salutary. The demise of organizations is on the whole traumatic for the people immediately involved. If many go under, the disturbance to the social system of which they are part may be widespread, prompting one well-known authority on organizational behavior to write: "In the past, innovation in society took place largely through the birth of new (innovative) organizations and the death of old (traditional) ones. Given the capital requirements of today's technology, this method seems a bit wasteful."[2] The death of organizations is more often regarded as regrettable and unfortunate and a sad necessity rather than as socially beneficial and desirable even by those who favor organizational mercy killings.[3] A theoretical model of organizations indicating that continuous replacement should be welcomed rather than deplored challenges that attitude and opens — or reopens — some neglected channels of inquiry.

The model, it will be recalled, rests on several premises. One is that organizations by and large are not capable of more than marginal changes, while the environment is so volatile that marginal changes are frequently insufficient to assure survival. Another is that organizational flexibility is ordinarily achieved at the cost of other advantages and thus does not necessarily increase the prospects for survival. Third is the deduction that replacement is therefore the principal process by which organizations adapted to their environment are formed. Fourth is the assumption that, in the absence of such a process of adaptation, most of humankind's organizations would eventually be overcome by environmental pressures, reducing humanity to an atomistic state in which its survival as a species would be in doubt. These are the reasons for maintaining that organizational turnover is a boon to humanity.

Turnover of this sort, I should add, implies commensurate variation among organizations. Obviously, if replacement organizations were identical to their predecessors in all respects, they would share the fate of their predecessors, so no evolution would take place. The process of adaptation operates not only through replacement of organizations by other organizations but by the appearance of successors different in at least some respects from those they replace. It requires the combination of variation and turnover.

Carried to the extreme, variation and turnover must turn into liabilities instead of assets. While turnover facilitates rapid adjustment to changing environmental conditions, if the organizations in any set are replaced by new ones at a rate approaching 100 percent in increasingly shorter intervals of time, the average life span of organizations in the continuously reconstituted set will decline because the young ones become almost the whole population. Under these circumstances the formation of new organizations would eventually be discouraged, for the gratifications of membership would be so fleeting that members would not be motivated to contribute their energies and other resources to them. Instead of enhancing adjustment capacity, extremely rapid turnover would become destructive.

Similarly, while variation assures such diversity that at least some organizations would be equipped for almost any environmental change imaginable, it could conceivably reach a level at which nothing is preserved or retained. The result would be ferment without development. So although the death of organizations and their replacement by younger variants are normally salubrious, the benefits would disappear if the levels climbed to the highest imaginable peaks.

Ordinarily, however, movement toward such an extreme would probably be arrested by the workings of natural selection before turnover and variation came anywhere near such dangerous levels. The trend would be self-limiting because the declining average life span of the organizational population would mean the ecological niches previously occupied by longer-lived organizations were being deserted, in which case some of the abundant variants suited to those zones, encountering no prior, established occupants, would take hold in them and repopulate them. As more longer-lived organizations appeared, turnover, which is a function of death rates as well as birth rates, would decline, and both average life span and rate of replacement would drift back toward moderate levels.

If the drifts should begin to approach zero turnover, on the other hand, which some commentators seem to favor, the danger of large-scale organizational collapse would once again threaten. The absence of young replacements would indicate that the resources of the medium in which organizations form had been absorbed almost entirely by aging organizations adapted to particular stable niches. In that event, environmental changes altering conditions in those niches would be fatal to them. Without an abundance of variants at hand to take their place, a long period of organizational disarray could conceivably ensue, to the discomfort and disadvantage of the people affected. Zero replacement in a world subject to the dynamics described by the hypothesis put forth here could be as dismaying as exceedingly high replacement.

Once again, however, self-correcting factors would come automatically into play. As older organizations succumbed to altered environmental circumstances for which their adjustment to their respective special ecological niches did not prepare them, the return of their contents—people, knowledge, skills, machinery, materiel, and so forth—to the medium in which organizations form would modify and enrich the medium, as postulated earlier.[4] The ecological zones once occupied by short-lived, swiftly evolving organizations, still sparsely populated because of the recent plethora of aged organizations, would now be open and fertile. This combination of conditions would make it possible for new organizations to take shape and for a higher rate of turnover to get started once again. Gradually, the tendency toward zero replacement would be checked and eventually reversed.

If these deductions are accurate, the turnover rate of organizations in any coherent human population should oscillate well within the extremes, although not necessarily around the midpoint between zero replacement and 100 percent replacement; in some populations the levels could oscillate around a higher level, in others around a lower one, depending on the carrying capacity, density, complexity, and other features of the particular society in which they occur. But in all cases oscillation would presumably occur as movement toward each pole triggered corrective tendencies.

Intuitively, I suspect that the antipodal thrusts toward opposite poles may not be symmetrical. The impulse toward low turnover seems to me to have a slight edge. In a stochastic world, that slight edge would result in an underlying drift toward the low end of the scale. My suspicion arises from the impression that established organizations often inhibit, by accident or design, the formation of new ones. At times this effect may be caused by nothing more than the entrenched units absorbing disproportionate amounts of needed resources from the surrounding medium. At other times the established ones may succeed, at least temporarily, in suppressing new ones. If my impression is right, turnover will "naturally" tend to decline rather than to rise, as will the degree of variation among organizations.

This reasoning suggests that humankind, to the extent that it has any influence at all on the course of events, has to guard more vigilantly against the dangers entailed by low turnover than by high rates of replacement. Instead of striving to *lower* organizational death rates and *extend* organizational longevity, we should, according to the implications of the evolutionary hypothesis, concentrate on *maintaining* birth rates and organizational diversity. The upward trend is less of a threat than the downward one.

In the United States, some public policies consistent with this strategy have actually been adopted—not, of course, because they are suggested by theory, but because of political pressures, ideological preferences for small entrepreneurs, and sympathy for presumed underdogs. Whatever the motives for the policies, they fit the logic of the evolutionary process. Assistance to small business entrepreneurs and family farmers and restraints on the use of economic power to prevent or suppress new entrants into old fields (including antitrust programs, prohibition of unfair competition, and proscription of discriminatory prices) all help shore up organizational birth rates that would otherwise fall.[5] These measures also contribute to organizational diversity, though perhaps not as much as do public support of research, experimentation, testing, and patent rights. Government policies on the distribution, sale, leasing, and licensing of public property and publicly regulated resources, as well as the policy of granting free access to many of the findings of government research undertakings and pilot and demonstration projects, likewise can be employed to foster new organizations and even new varieties of organization.[6] The possibilities are legion.

The logic underlying public policy could conceivably be applied by organizations smaller than governments; there is no self-evident reason why it should not be. Planning groups and research and development teams intended to foster innovative policies and products are already common in most large-scale organizations, both public and private. Perhaps managers could also experiment with techniques of stimulating new units and methods and practices within their structures. What is beneficial at a societal level might well be advantageous at other levels. Maintaining turnover among the components of organizations (i.e., the organizations within organizations) by keeping up internal organizational birth rates would seem to be as strongly indicated for the one as for the other.

Nobody can *guarantee,* however, that the policies suggested here will produce the results intended—or any results at all—at any organizational level. If the process of organizational evolution is truly governed by its own dynamics, a dynamics in which human purposes and wishes play a circumscribed part and chance plays a very large role, then the prospects for human control of the process are very dim. Like all models of human systems based on the driving force of assumed impersonal factors, this one makes human striving seem irrelevant and pointless. If what happens is beyond our capacity to influence it in any significant way—if, indeed, our efforts may be followed by the opposite of what we want—why should anyone labor to direct the course of events?

I am convinced there *are* reasons for doing so. The main one is that humankind as a whole learns from its endeavors to manipulate the world. Much is learned from failures as well as from successes. If we stop trying, we also stop learning. While I think it unlikely that we will acquire enough knowledge in the near future to make organizational evolution an instrument of human will, fathoming its inner workings may enable us to avoid a few costly errors and escape a few futile undertakings. Moreover, there is pleasure in the quest for knowledge and in the occasional moments of discovery. Why surrender these rewards just because we cannot assure ourselves of total mastery of the process?[7]

In any case, even if someone produced clear proof that we could *never* bring the global process under our control, there is still the possibility that we might discover in it some ways to make things a little more pleasant for ourselves at the level of the individual. Just as the birth or death of a particular person has little impact on statistical birth and death rates but is of major significance to the person involved and those close to him or her, so our attempts to make the organizational world do our bidding may be trivial on the grand scale but may alter our own lives and the lives of those around us in ways important to us. The more we understand about the process, it seems to me, the higher the probability that we can make at least small gains of this kind. Even if they are only brief respites from the frustrations and feelings of helplessness that afflict so many of us so much of the time, they would be worth working for.

Perhaps we can even accumulate sufficient understanding of the process to let us capitalize on its thrust the way a sailing vessel makes use of the wind. Once we grasp its principles we might be able to ride with it where we want to go — or at least avoid being swept where we don't want to go. We don't have to command it in order to turn it to our ends; even if it is never subjugated, we can work with it, draw on it, make it a resource instead of a burden.

If organizational evolution does flow like a current, the first step toward reaping its benefits while minimizing its disadvantages is to acknowledge its existence. If we insist that organizations are nothing more than human artifacts that we can fashion and maneuver according to our tastes, we are bound to be buffeted by the current's force as we hurl ourselves blindly into it or against it. We have to recognize it before we can take full advantage of it.

To profit from the current of organizational evolution we shall have to study it systematically. My portrait of it is only a first approximation, so I would not be surprised if this formulation turns out to be a very crude likeness. I believe it will improve as we methodically gather more information about it. The concluding chapter, the Postscript, is a discussion of the kinds of information I think we need.

NOTES

1. See note 6.
2. Victor A. Thompson, *Bureaucracy and Innovation* (University, Ala.: University of Alabama Press, 1969), 4.
3. A notable exception was Joseph A. Schumpeter, who described capitalist economic development as ceaseless revolutionizing of the economic structure by "industrial mutation," which is "incessantly destroying the old one, incessantly creating a new one," a process he dubbed "Creative

Destruction"; *Capitalism, Socialism, and Democracy* (New York: Harper, 1942), chap. 7, esp. p. 83.

4. See pages 91-96.

5. Of course, contrary policies, including legal barriers to foreign and domestic competition, and subsidies and loans and loan guarantees to established firms, tend to extend the longevity of some organizations; the thrust is not all in one direction. My point is merely that promoting higher birth rates and diversity is an actual, not merely a theoretical, possibility, and has been attempted. The logic of the evolutionary process indicates that shifting the balance of public policies toward this pole, if achieved, could have significant salutary results for the system.

6. There were also attempts to promote legislation intended to increase the death rate among government agencies — so-called Sunset Legislation — by limiting the duration of their enabling statutes to fixed periods unless renewed by law; Bruce Adams, "Sunset: A Proposal for Accountable Government," *Administrative Law Review* 28, no. 3 (Summer 1976): 511-40; U.S. Senate, *Sunset Act of 1977, Hearings before the Subcommittee on Intergovernmental Relations of the Senate Committee on Governmental Affairs*, 95th Cong., 1st sess, 1979. The practice of limited duration could also be applied (and, indeed, occasionally has been applied) to licenses and charters granted to nongovernmental organizations to conduct activities of various kinds. One effect of such measures might be an increase in organizational turnover. Destroying organizations, however, is more difficult and painful than encouraging them. The movement to deliberately raise death rates apparently petered out. Still, it certainly remains as a possible strategy for elevating rates of organizational replacement.

7. Maybe we should dare to explore organizational evolution, and feel comfortable about doing so, *because* it appears we cannot control the pace; our lack of control, after all, should give us confidence that we are not capable of making things significantly worse than they otherwise would be.

Q. How Can We Find Out if the
Hypothesis Is Right?

A. By Trying to Prove It Wrong.

Hypotheses that cannot be tested for possible falsity by any feasible or potentially feasible empirical study lie beyond the realm of science. They may have much else to commend them, but they are not science.[1]

My hypothesis about organizational evolution is testable. The test, unfortunately, would not be fast or easy. With enough will and resources and time, however, it is practicable.

The Threshold Test:
Does the Organizational Medium "Thicken"?

According to the hypothesis, organizations form out of a medium consisting of people, culture in the fullest sense, and energy; and the medium is enriched, or "thickens," as a result of the activities of organizations—even when organizations dissolve, for their contents return to the medium and are recycled by other organizations.[2] Therefore, if we could measure the dimensions of "thickness," we could discover whether the hypothesis is false. If the medium does not change as the hypothesis indicates it must, something must be wrong with the hypothesis.

That is not to say it is "proven" or that all other possible explanations are ruled out if the medium *does* change as the hypothesis implies. After all, the "thickening" of the medium might be accounted for by factors other than, or at least in addition to, organizational variation and natural selection; other theories might be consistent with the same data. An ideal decisive test would presumably select among them.

Unfortunately, formulating the hypothesis precisely enough to permit such finality and contriving a single instrument for this purpose, even if such ideals are attainable in this field, exceed my powers. All I can propose is a line of inquiry that will expose the error of my hypothesis if it is wrong, not one that will also eliminate all conceivable alternatives. Any contest between the evolutionary hypothesis and other explanations of observed trends in the character of the organizational medium will have to be resolved by researches designed specifically to compare the validity and explanatory power of each other proposal with the one put forth here. No test I can conceive of will do everything in one ingenious swoop.

I believe that my hypothesis, if it is not refuted by the basic test described below, will stand up well when tested against specific competitors. The reason for this belief is that everything I can imagine that might contribute to the "thickening" of the organizational medium seems to occur within organizations, or as a result of relations among organizations, or in consequence of the activities of organizations; other contributory factors emerge and work their effects only through organizations, and their potency looks as though it is somehow linked to the number and interdependence and complexity of the organizations in which they appear. I therefore anticipate that organizational evolution will explain more of what we observe, and explain it more thoroughly and dependably, than other models also consistent with trends in the medium. My impression may be proved wrong by empirical research, of course; it is only a hunch at this time. But it is not groundless.

We will never have to explore this possibility, however, if the medium does *not* behave as my hypothesis predicts it must. That is, if the medium does not "thicken," neither my hypothesis nor any other explanation of why it must "thicken" makes sense. Hence, before we get to the problem of selecting among competing ideas, we must first find out whether the predicted effects occur; that is the threshold test. And that test, I submit, is well within our present capabilities.

THE MEANING OF "THICKNESS"

When I say organizations form out of the medium of people, culture, and energy, I am simply paralleling the accepted notion that living things formed out of the "dust" of the planet. Organizations are not the same as living things because the medium out of which they form is different. But the processes by which the two kinds of structures arise are similar in that the structures are products of conjunctions of circumstances that bring elements of their respective media together in eventually self-sustaining formations.

When I say the medium out of which living things formed became thicker, I mean the number and complexity of the ingredients comprised in it increased with time. Similarly, when I say the medium out of which organizations form becomes thicker, I mean that the activities of people and organizations become increasingly differentiated, with the result that interdependencies multiply and levels of knowledge and skills rise. These developments in turn intensify differentiation so that more organizations are propagated, and the process spirals upward.

In any brief interval of time in any given locality the reverse may take place. People may revert to earlier patterns, differentiation may decline, interdependencies may decrease, and the medium may thin. We have evidence from the remnants of artifacts, after all, that some communities dwindled and disappeared.

Therefore it would be folly to claim that the medium in which organizations form grows thicker everywhere and constantly. Like other natural processes, this one is not uniform or steady.

That means the index must be the *average* thickness of the medium *globally;* otherwise there is no way to test the hypothesis. That is, the hypothesis permits us to predict only a *net* upward rise in its thickness throughout the world. Locally, contrary tendencies may occur, but if the hypothesis is right they will be offset by thickening elsewhere — and even the local tendencies will be reversed as the thickening elements diffuse through the entire medium.

Conceptually, that is enough to permit a test. If studies should turn up a *single* instance of a decline in the average thickness of the global medium, the hypothesis would have to be rejected. It stands or falls on the validity of this postulated trend.

Obviously, measuring the thickness of the medium on a global scale would be a monumental task. Is it beyond the limits of feasibility? Is the proposed test of the hypothesis a sophism because there is no hope of conducting it? Is the appearance of rigor a sham?

I think not. Conducting the test would clearly require great ingenuity, resolve, perseverance, patience, and money. Whether it would merit the effort is not a question everyone will answer affirmatively; perhaps no one will. But that is a different issue. A test may be practicable even if nobody regards it as worthy of doing. Whatever the attractiveness of the hypothesis proposed here, *testing* it is not inherently impossible.

MEASURING THICKNESS

The thickness of the medium can be represented by seven variables. I do not mean to imply that they measure every attribute of the medium; they do not constitute a full, detailed portrait of it. Rather, they are qualities that, for reasons I shall offer, seem to me useful as crude indexes of its thickness. And the tests do not demand highly refined measures; the hypothesis implies only the *direction* of development, not the magnitudes or speed. The indicators, therefore, need not be elaborate.

The ones I propose are (1) organizational specialization, (2) occupational specialization, (3) literacy and educational levels, (4) cultural institutions and personnel, (5) volume and speed of communication, (6) energy consumption per capita, and (7) organizational density. Taken together, they indicate how interdependent people and organizations are, how much knowledge and skill are accessible and are employed, how high the level of technology is likely to be, and how intense will be the competition among organizations for the available pool of human energy. The higher these indicators, the more hospitable the environment to the formation of organizations of many kinds and the greater the possible complexity of at least some of the organizations that appear. Let me explain this contention by looking at each of the indicators individually.

Organizational specialization refers to the division of labor among organizations. If one were to imagine a world in which there was an abundance of organizations no two of which performed the same vital functions, the degrees of interdependence and exchange among them would be extremely high. In such a setting the need for mediating organizations, coordinating groups, and communication and transportation bodies would be very strong, and they would form in profusion. The appearance of these organizations would further complicate the system and intensify the need for still more specialists and organizational layers. A system like that would be so precarious because of the absence of redundancy that it would not last long; it is a theoretically limiting case rather than a significantly probable occurrence. But the theoretically extreme case dramatizes the principle that organizational specialization multiplies relationships and spawns organizations, and in this sense indicates a "thickening" of the organizational medium.

For many of the same reasons, so does occupational specialization by individuals. Moreover, since specialized organizations would be likely to encompass numbers of identical occupational specialties, even though the outputs of the organizations are different, many people would be able to move from one organization to another without difficulty. Varieties of personal organizational experience would thus be joined to their occupational expertise, creating new combinations of skill and experience and engendering new activities, organizations, and capabilities. These qualities, being internalized within individuals, would become elements of the medium independent of the organizations in which they were acquired, thereby enriching the medium and the organizations that subsequently form in it.

As the skills and knowledge needed to operate organizations rise, specialized institutions providing instruction in basic information, values, leadership, technical knowledge, and professional expertise would increase as well. Thus the extent of literacy, the number of schools and other instructional institutions, the number of people attending them, and the number of personnel operating them (all standardized in some fashion) constitute a rough clue to the kind of system the people inhabit and the kinds of organizational life they can support. In very highly developed systems, museums, research centers, libraries, and other methods of storing, increasing, retrieving, and sharing the collective knowledge of the population multiply, and the means of disseminating such information accumulate. Hence the third, fourth, and fifth indicators.

Energy (not including human energy) consumption per capita I take to be a mark of technological advancement. Where the rate is very high, the organizational environment is likely to be compact and crowded, and the medium for organizations, I suspect, extraordinarily fertile; where it is low, the environment is likely to be sparse and barren. That accounts for adoption of this item as the sixth indicator.

The seventh, organizational density—that is, the number of organizations per hundred (or other convenient unit) of human population—is the most straightforward and direct of the indicators. Obviously, the higher this ratio, the greater the volume of organizations and the chances that complex combinations of organizations will form. When the medium is rich in organizations, it is probably especially likely to propagate new levels of complexity.

It would not be surprising if all these indicators moved more or less in tandem; indeed, it would probably be surprising if they were wholly independent. Should that prove to be the case, we could dispense with some or even most of them, using just one or two as our index of the richness of the medium. That would certainly simplify the task of measurement.

I would not want to infer a priori that they *will* move together in the same direction, however. Although the logic of their selection and the way they are defined seems to make this outcome inevitable, I would not feel comfortable in assuming this will be the case until analysis of empirical data furnishes a solid foundation for such an inference. They *might* change at markedly different rates, and some might even decline when others are going up. We ought to know before we take steps to simplify our measurements.

If they do behave differently, the problem of assigning weights and blending them into a combined index will be most challenging. And even if they turn out to be so alike that any one of them could be employed as a surrogate for all of them collectively, it will not be easy to invent techniques of averaging differences in rates and direction of change in different parts of the global human population. Under the most favorable circumstances there would be a long way to go, and we cannot count on the circumstances being favorable.

On the other hand, we do not have to get off from a standing start. The amount of information already being collected is gratifying,[3] and under the auspices of the United Nations, much of it is worldwide.[4] The detail is not equally complete on all subjects and for all areas; we would have to begin from scratch, for example, on assessing organizational density. Still, although protracted work is inescapable, there is a good foundation on which to build.

Despite the foundation, I would be disinclined to try to reconstruct the entire history of the medium's thickness. Only recently have sound, inclusive international records been kept. It makes more sense to start tracking the indicators from the present (or very recent past) forward into the future. We will have a long wait for probative results, but that is better than succumbing to the temptation to seek quick outcomes by applying indexes to earlier periods for which data are sparse and uncertain. When the techniques of measurement are perfected, perhaps they can be employed retrospectively. Meanwhile, it seems to me that we should start where we are and go forward.

Another temptation such a project will face is the likely predisposition on the part of many people to pronounce judgments on the dimensions of the organizational medium without laborious research. A few anecdotes, sweeping characterizations of whole epochs of human history, and other impressionistic generalizations will be advanced as grounds for dispensing with painstaking investigation. Ultimately, impressionistic conclusions may prove right. Until we have a reliable basis for deciding that they are, however, I will remain skeptical of any impressions not backed by solid data, no matter how "self-evident" they are said to be.

There is a substantial probability that the painfully gathered and analyzed data will prove the hypothesis wrong, a risk that should not be minimized by anyone undertaking the research. The risk does not mean the research should not be done; the only research worth doing is that whose results are not known in advance. The justification for taking such chances is not only the hope that the hypothesis will survive its test, but the conviction that disproving erroneous hypotheses is a useful endeavor and that, in any case, we often advance knowledge in the course of conducting the test. I would not let the possibility that the outcome may run counter to my wishes deter me from the attempt.

The Road Beyond

If the decisive test *is* conducted and if the hypothesis *does* survive, the study of organizational evolution would not then end. On the contrary, it would just have begun. For one thing, the decisive test ought to be repeated many times to assure that results consistent with the hypothesis were not mere flukes. For another, if the hypothesis withstands the test, it will open many additional lines of inquiry.

The additional lines of inquiry stem from unproved assumptions about properties of organizations and their environment presented in the argument put forth in the preceding pages, and also from unexamined relationships implicit in the analysis, that clamor for scrutiny. The hypothesis would not necessarily be disproved if some individual guesses are shown to be wrong or if unexpected correlations are discovered; conclusions may stand even if some of the steps on the way to them are off-target. Nonetheless, correcting such errors would doubtless improve our understanding of organizational evolution and lead to refinements of the hypothesis that my first formulation of it cannot achieve.

For example, the connection between organizational turnover and the thickening of the organizational medium calls for clarification. High turnover, which is a product of high birth and death rates, would seem to contribute to increased thickening by speeding the process of evolution. But perhaps a high organizational birth rate combined with a low death rate, and a steady rise in the number of extremely long-lived (immortal?) organizations, would also produce the same effect. Indeed, is it possible that the medium can become so dense, so viscous, that it jells, congeals, and slows evolution?[5] Does it vary from place to place and over time? If so, why, and with what consequences? Can public policy affect the medium in predictable and controllable fashion? Without solid data on organizational births, deaths, and longevity, as well as on the medium, we can only speculate.

Similarly, my proffered explanation of those apparently rare cases of extremely long organizational life calls for investigation. *Are* there stable ecological niches in the midst of all the environmental turbulence besetting the organizational environment? Can we identify the critical elements of the situations that enable long-lived organizations to survive for such protracted periods while most of their sister organizations succumb to swirling change? What are the effects of different kinds of environments on organizations?[6] Are organizational mortality and environmental volatility, as postulated, strongly associated with each other? Are there, as I have suggested,[7] some sorts of built-in dampers that terminate organizational responses to uncertainty before they reach destructive extremes? Is imperfect execution of organizational decisions (rather than the substance of the decisions themselves) as significant a cause of organizational demise as I predict?[8] Do differences in intelligence and ability count for as little as I have postulated[9] in the survival and extinction of organizations? Our understanding of organizational evolution would surely be enhanced by accurate and reliable answers to these questions.

It might be improved also by a taxonomy of organizations linked to the evolutionary schema. We do not lack for classifications of organizations based on structure, function, distribution of power and benefits, form of ownership, degree of governmental control, size, age, and other criteria, each of which serves the particular purposes of different theorists.[10] Such diversity seems to me altogether fitting; no single principle of classification is likely to satisfy differing goals. What we are still groping toward, however, is a taxonomy that takes account of the developmental implications of evolution and calls attention to connections among the categories as well as to the distinctions between them. Evolution implies a progression in the history of organizations, but few of the prevailing classifications have been designed in a way that would display the progression even if it could be detected.[11]

Another part of the argument in need of attention is my low estimate of the importance of leadership in the evolution of organizations. I contended that the Tolstoyan view of leaders as chips tossed about by the tides of history rather than as masters of events cannot be rejected a priori, and I speculated that this proposition about the role of leadership in organizations could be systematically examined if our presuppositions about leadership were set aside for a time while we looked at it dispassionately. I also suggested that even if leaders *do* appear to be as important as conventional opinions hold them to be, the quality of leadership will nevertheless prove to be randomly rather than systematically distributed among organizations, and chance will therefore remain the main factor in organizational survival. This statement, too, can be researched empirically. Contriving a method of assessing the quality of leaders that does not employ organizational survival itself as the determining measure is a challenging conceptual and research problem. If it can be done, another of the assertions in my argument can be objectively evaluated and corrected as necessary.

In like fashion, what I said about the costs and benefits of organizational flexibility in the survival of organizations, about organizational complexity, and about the effects of strong and weak bonding of members to one another and to their organizations, could be explored if someone invented methods of assessing flexibility and complexity and unity. (Case studies of the deaths of organizations would enrich these inquiries.) In addition, human populations characterized by high organizational density should be compared with those of low density in order to study the causes and effects of this variable. It would be especially interesting to test Chester I. Barnard's intriguing speculation that there might be more organizations than people in the world.[12] And I should be curious to find out whether my ruminations on the tendencies of organizational turnover and the possibilities of influencing the level through public policies have any validity.

This list of researchable questions stimulated by the evolutionary hypothesis does not, I suspect, include all the assertions and assumptions in or underlying my argument. It certainly does not exhaust the array of deductions that may be drawn from the model. It does, however, give some idea of the abundance of propositions imbedded in or flowing from the hypothesis. It explains why I maintain that subjecting the hypothesis to the decisive test would constitute only the beginning of research implied by the logic of organizational evolution.

Some of these questions, of course, occur to students in the absence of the hypothesis and can be investigated regardless of its fate. Indeed, some of them already have been targets of inquiry and experimentation, thanks to the experience, intuition, and lively curiosity of students of organization. Still the hypothesis calls attention to areas of research that probably would not otherwise become salient. It also relates them to one another within the framework of an overarching theoretical concept, which can add to their richness and explanatory power. At any rate, the hypothesis presents a full platter of items for research, not just a single dish.

Testing the hypothesis, refining it, and tracing out its implications call for great imagination, inventiveness, and persistence. The initial step of formulating it may well, in retrospect, seem easy by comparison. As Aesop told us, any mouse can recognize the principle that putting a bell on a predatory cat would enable mice to escape their nemesis; *executing* the strategy demands extraordinary gifts. In the evolution of the theory of organizational evolution, my hat will be off to those who succeed in belling the empirical cat.

NOTES

1. Karl R. Popper, *The Logic of Scientific Discovery* (New York: Basic Books, 1959), 32-34, 78-92.
2. See pages 91-93, 94-96.

The same logic might apply generally to the whole biosphere as well as to the human component of that layer of living things. The biosphere is packed tight with life. "Credit is given to Thomas Henry Huxley for an analogy for the filling of the earth with life. He likens it to the filling of a barrel with apples until they heap over the brim. Still there is space into which quantities of pebbles fit before they overflow. Again sand is added, and much of it packs down between the apples and the pebbles. The barrel is not yet full and quarts of water may be poured in before at last the barrel can hold no more"; George Gaylord Simpson, *The Meaning of Evolution: A Study of the History of Life and of Its Significance for Man* (New Haven: Yale University Press, 1949), 112-13. If the water in the metaphor were replaced by oil that becomes even more viscous with the passage of time, the parallel with my image of the organizational medium would be very close indeed.

3. For a comprehensive overview of sources, see Paul Wasserman, Jacqueline O'Brien, and Kenneth Clansky, eds., *Statistics Sources: A Subject Guide to Data on Industrial, Business, Social, Educational, Financial, and Other Topics for the United States and Internationally,* 6th ed. (Detroit: Gale, 1980), esp. 1-17.

See also U.S. Department of Commerce, Office of Federal Statistical Policy and Standards, *Standard Industrial Classification Manual* (Washington, D.C.: Government Printing Office, 1972; *Supplement,* 1977); *Enterprise Standard Industrial Classification Manual* (Washington, D.C.: Government Printing Office, 1974); *Standard Occupational Classification Manual* (Washington, D.C.: Government Printing Office, 1977), for illustrations of the raw materials available for the United States. (These documents were formerly produced by the U.S. Office of Management and Budget, Statistical Policy Division.) Corresponding data for other countries have been published by the United Nations; see note 4.

Additional projected materials for the United States have been advocated by components of the American Academy of Arts and Sciences. See Raymond A. Bauer, ed., *Social Indicators* (Cambridge: MIT Press, 1966); and the Social Science Research Council *Annual Report, 1980-81,* 118-24.

For data on political parties around the world, see Kenneth Janda, *Political Parties: A Cross-National Survey* (New York: Free Press, 1980).

4. For a full list of materials put out by the United Nations and its specialized

agencies, see *UNDOC: Current Index* (New York: United Nations, ten issues per year plus annual accumulations). The volume of data is imposing.
5. This may be the implication of the works cited in the second paragraph of note 36 in chapter 3.
6. Howard E. Aldrich, in chap. 3 of his *Organizations and Environments* (Englewood Cliffs, N.J.: Prentice-Hall, 1979), sets forth a classification of environments that could serve as a point of departure for research on this question.
7. See pages 43-44.
8. See pages 51-53.
9. See pages 69-71.
10. For a review and discussion of typologies of organizations, see W. Richard Scott, *Organizations: Rational, Natural, and Open Systems* (Englewood Cliffs, N.J.: Prentice-Hall, 1981), chap. 2.

    See also the functional classification by Luther Gulick in "Notes on the Theory of Organization" and by L. Urwick, "Organization as a Technical Problem," *Papers on the Science of Administration,* ed. Luther Gulick and L. Urwick (New York: Institute of Public Administration, 1937), 15-30, 49-88.

    For a structural criterion, see Herbert A. Simon, Donald W. Smithburg, and Victor A. Thompson, *Public Administration* (New York: Knopf, 1950), 268-72.

    James Q. Wilson employed incentives offered members of organizations as a principle of classification in *Political Organizations* (New York: Basic Books, 1973), chap. 3.

    Other common (and dichotomous) categories include public versus private, regulated versus unregulated, large versus small, old versus young, complex versus simple, and autocratic versus democratic organizations.
11. Notable exceptions are the efforts of Bill McKelvey, whose work is cited and summarized in Scott, *Organizations,* at 51-53; and of John Freeman and Michael T. Hannan in their "Niche Width and the Dynamics of Organizational Populations," *American Journal of Sociology* 88, no. 6 (May 1983): 1116-45.
12. Chester I. Barnard, *The Functions of the Executive* (Cambridge: Harvard University Press, 1938), 4.

    William N. Parker, of the Department of Economics at Yale University, showed me, some two decades before the appearance of this book, that Barnard's proposition could be tested for a specific human population by sampling. His logic was as follows:

    If n = the number of people in a given population, and
    q = the average number of organizations to which a
    person belongs, and

S = the average size of an organization in that population, and
N = the number of organizations in that population, then

$n\, q/S = N,$ or

$q/S = N/n$

If N is ever greater than n, q must in such cases be greater than S. We can sample for values of q and S by selecting a random or stratified subset of the entire human population and compiling the total number of organizations to which they belong. Dividing the total number of memberships by the number of people in the sample yields q; dividing the total number of memberships by the total number of organizations turned up in the sample yields S.

Select Bibliography

Select Bibliography

ALCHIAN, ARMEN A. "Uncertainty, Evolution, and Economic Theory." *Journal of Political Economy* 58, no. 3 (June 1950): 211-21.

ALDRICH, HOWARD E. *Organizations and Environments.* Englewood, Cliffs, N.J.: Prentice-Hall, 1979.

ALEKSANDER, IGOR, and BURNETT, PIERS. *Reinventing Man: The Robot Becomes Reality.* New York: Holt, Rinehart and Winston, 1983.

AMBROSE, E. J. *The Nature and Origin of the Biological World.* Chichester [West Sussex], England: Ellis Horwood Limited, 1982.

ASHBY, W. ROSS. "The Self-Reproducing System." In *Aspects of the Theory of Artificial Intelligence: The Proceedings of the First International Symposium on Biosimulation, Locarno, 1960,* edited by C. A. Muses, 9-18. New York: Plenum Press, 1962.

AYRES, ROBERT U., and MILLER, STEVEN M. *Robotics: Applications and Social Implications.* Cambridge: Ballinger, 1983.

BAKKE, E. WIGHT. *Bonds of Organization: An Appraisal of Corporate Human Relations.* New York: Harper, 1950.

BARNARD, CHESTER I. *The Functions of the Executive.* Cambridge: Harvard University Press, 1938.

BERNSTEIN, MARVER H. *Regulating Business by Independent Commission.* Princeton: Princeton University Press, 1955.

BLAU, PETER M., and SCOTT, W. RICHARD. *Formal Organizations: A Comparative Approach.* San Francisco: Chandler, 1962.

BONNER, JOHN TYLER. *Cells and Societies.* Princeton: Princeton University Press, 1955.

BORKO, HAROLD, ed. *Computer Applications in the Social Sciences.* Englewood Cliffs, N.J.: Prentice-Hall, 1962.

BOULDING, KENNETH E. *Ecodynamics: A New Theory of Social Evolution.* Beverly Hills, Calif.: Sage, 1978.

BOULDING, KENNETH E. *The Organizational Revolution: A Study in the Ethics of Economic Organization.* New York: Harper, 1953.

BROOKS, HARVEY. "Technology, Evolution, and Purpose." In *Science, Technology, and National Purpose,* edited by Thomas J. Kuehn and Alan L. Porter, 35-57. Ithaca, N.Y.: Cornell University Press, 1981.

BURNS, JAMES MACGREGOR. *The Deadlock of Democracy: Four-Party Politics in America.* Englewood Cliffs, N.J.: Prentice-Hall, 1963.

BURNS, JAMES MACGREGOR. *Leadership.* New York: Harper & Row, 1978.

CAIRNS-SMITH, A. G. *Genetic Takeover and the Mineral Origins of Life.* New York: Cambridge University Press, 1982.

CAMPBELL, DONALD T. "Common Fate, Similarity, and Other Indices of the Status of Aggregates of Persons as Social Entities." *Behavioral Science* 3, no. 1 (January 1958): 14-25.

CAMPBELL, DONALD T. "On the Conflicts Between Biological and Social Evolution and Between Psychology and Moral Tradition." *American Psychologist* 30, no. 12 (December 1975): 1103-26; and "Reprise," *American Psychologist* 31, no. 5 (May 1976): 381-84.

CAMPBELL, DONALD T. "The Two Distinct Routes Beyond Kin Selection to Ultrasociality: Implications for the Humanities and Social Sciences." In *The Nature of Prosocial Development: Theories and Strategies,* ed. Diane Bridgeman, 11-41. New York: Academic Press, 1983.

CAMPBELL, DONALD T. "Variation and Selective Retention in Socio-Cultural Evolution." In *Social Change in Developing Nations: A Reinterpretation of Evolutionary Theory,* edited by Herbert R. Barringer, George I. Blanksten, and Raymond W. Mack, 19-48. Cambridge: Schenkman, 1965.

CANNON, WALTER B. *The Wisdom of the Body.* Revised ed. New York: Norton, 1939.

CARROLL, GLENN R., and DELACROIX, JACQUES. "Organizational Mortality in the Newspaper Industries of Argentina and Ireland: An Ecological Approach." *Administrative Science Quarterly 27*, no. 2 (1982): 169-98.

CARTWRIGHT, DORWIN. *"Influence, Leadership, Control."* In *Handbook of Organizations*, edited by James G. March, 1-47. Skokie, Ill.: Rand McNally, 1965.

CASSTEVENS, THOMAS W. "Population Dynamics of Governmental Bureaus." *The UMAP Journal* 5 (1984): 178-99.

CHAMBERLAIN, NEIL W. *Enterprise and Environment: The Firm in Time and Place.* New York: McGraw-Hill, 1968.

COHEN, MICHAEL D.; MARCH, JAMES G.; and OLSEN, JOHAN P. "A Garbage Can Model of Organizational Choice." *Administrative Science Quarterly* 17 (1972): 1-25.

COKER, FRANCIS W. *Organismic Theories of the State: Nineteenth Century Interpretations of the State as Organism or as Person.* New York: Columbia University Press, 1910.

CORNING, PETER A. *The Synergism Hypothesis: A Theory of Progressive Complexity.* New York: McGraw-Hill, 1983.

CROZIER, MICHEL. *The Stalled Society.* New York: Viking, 1974.

DARLINGTON, C. D. *The Evolution of Man and Society.* New York: Simon and Schuster, 1969.

DAWKINS, RICHARD. *The Selfish Gene.* New York: Oxford University Press, 1976.

DELACROIX, JACQUES, and CARROLL, GLENN R. "Organizational Foundings: An Ecological Study of the Newspaper Industries of Argentina and Ireland." *Administrative Science Quarterly* 28, no. 2 (June 1983): 274-91.

DIEBOLD, JOHN, ed. *The World of the Computer.* New York: Random House, 1973.

DOBZHANSKY, THEODOSIUS. *Mankind Evolving: The Evolution of the Human Species.* New Haven: Yale University Press, 1962.

DOWNS, ANTHONY. *Inside Bureaucracy.* Boston: Little, Brown, 1967.

DOWNS, ANTHONY. *Neighborhoods and Urban Development.* Washington, D.C.: Brookings Institution, 1981.

DUBOS, RENE. *Man Adapting.* New Haven: Yale University Press, 1965.

DUN AND BRADSTREET. *Business Failure Record, 1982-1983.* New York: 1985.

DUNN, EDGAR S., JR. *Social and Economic Development: A Process of Social Learning.* Baltimore: Johns Hopkins University Press, 1971.

EISELEY, LOREN. *The Unexpected Universe.* New York: Harcourt, Brace, and World, 1969.

FEIGENBAUM, EDWARD A., and FELDMAN, JULIAN, eds. *Computers and Thought.* New York: McGraw-Hill, 1963.

FELDMAN, JULIAN, and KANTER, HERSCHEL E. "Organizational Decision Making." In *Handbook of Organizations,* edited by James G. March, 614-49. Skokie, Ill.: Rand McNally, 1965.

FESLER, JAMES W. *Public Administration: Theory and Practice.* Englewood Cliffs, N.J.: Prentice-Hall, 1980.

FOX, SYDNEY W., and DOSE, KLAUS. *Molecular Evolution and the Origins of Life.* Revised ed. New York: Marcel Dekker, 1977.

FREEMAN, JOHN, and HANNAN, MICHAEL T. "Niche Width and the Dynamics of Organizational Populations." *American Journal of Sociology* 88, no. 6 (May 1983): 1116-45.

GALBRAITH, JOHN KENNETH. *American Capitalism: The Concept of Countervailing Power.* Boston: Houghton Mifflin, 1952.

GOLEMBIEWSKI, ROBERT T. *Renewing Organizations: The Laboratory Approach to Planned Change.* Itasca, Ill.: Peacock, 1972.

GOODMAN, PAUL S., et al. *Change in Organizations: New Perspectives on Theory, Research, and Practice.* San Francisco: Jossey-Bass, 1982.

GORDON, ROBERT AARON. *Business Leadership in the Large Corporation.* Berkeley: University of California Press, 1961.

GOULD, STEPHEN JAY. *The Panda's Thumb: More Reflections in Natural History.* New York: Norton, 1982.

GOULDNER, ALVIN W., ed. *Studies in Leadership: Leadership and Democratic Action.* New York: Harper, 1950.

GROSS, BERTRAM M. *The Managing of Organizations: The Administrative Struggle.* New York: Free Press, 1964.

GUEST, ROBERT H. *Organizational Change: The Effect of Successful Leadership.* Homewood, Ill.: Dorsey Press, 1962.

GULICK, LUTHER. "Notes on the Theory of Organization." In *Papers on the Science of Administration,* edited by Luther Gulick and L. Urwick, 1-45. New York: Institute of Public Administration, 1937.

HALPERIN, MORTON H. *Bureaucratic Politics and Foreign Policy.* Washington, D.C.: Brookings Institution, 1974.

HANNAN, MICHAEL T., and FREEMAN, JOHN. "The Population Ecology of Organizations." *American Journal of Sociology* 82, no. 5 (March 1977): 929-64.

HASKINS, CARYL P. *Of Societies and Men.* London: George Allen and Unwin, 1952.

HAYEK, FRIEDRICH A. *Law, Legislation and Liberty.* Vol. 1, *Rules and Order.* Chicago: University of Chicago Press, 1973.

HAYEK, FRIEDRICH A. *The Three Sources of Human Values.* London: London School of Economic and Political Science, 1978.

HAYNES, W. WARREN, and MASSIE, JOSEPH L. *Management: Analysis, Concepts, and Cases.* 2d ed. Englewood Cliffs, N.J.: Prentice-Hall, 1969.

HEDBERG, BO. "How Organizations Learn and Unlearn." In *Handbook of Organizational Design,* Vol. 1, *Adapting Organizations to Their Environments,* edited by Paul C. Nystrom and William H. Starbuck, 3-27. New York: Oxford University Press, 1981.

HIRSHLEIFER, JACK. "Economics from a Biological Viewpoint." *Journal of Law and Economics,* 20, no. 1 (April 1977): 1-52.

HOAGLAND, MAHLON B. *The Roots of Life: A Layman's Guide to Genes, Evolution, and the Ways of Cells.* Boston: Houghton, Mifflin, 1979.

HOFSTADTER, RICHARD. *Social Darwinism in American Thought.* Revised ed. Boston: Beacon Press, 1955.

HOMANS, GEORGE C. *The Human Group.* New York: Harcourt, Brace, and World, 1950.

HOOK, SIDNEY. *The Hero in History: A Study in Limitation and Possibility.* Reprint ed. Atlantic Highlands, N.J.: Humanities Press, 1950.

HUNT, H. ALLAN, and HUNT, TIMOTHY L. *Human Resource Implications of Robotics.* Kalamazoo, Mich.: W. E. Upjohn Institute for Employment Research, 1983.

HUXLEY, ALDOUS. *Brave New World.* Garden City, N.Y.: Doubleday, 1932.

JASTROW, ROBERT. *The Enchanted Loom: Mind in the Universe.* Austin, Tex.: S & S Press, 1981.

KAPLAN, LAWRENCE. "Life Span." *The New Encyclopaedia Britannica.* 15th ed. 1982.

KAUFMAN, HERBERT. *The Administrative Behavior of Federal Bureau Chiefs.* Washington, D.C.: Brookings Institution, 1981.

KAUFMAN, HERBERT. *Are Government Organizations Immortal?* Washington, D.C.: Brookings Institution, 1976.

KAUFMAN, HERBERT. *The Forest Ranger: A Study in Administrative Behavior.* Baltimore: Johns Hopkins University Press, 1960.

KAUFMAN, HERBERT. *The Limits of Organizational Change.* University, Ala.: University of Alabama Press, 1971.

KAUFMAN, HERBERT. *Red Tape: Its Origins, Uses, and Abuses.* Washington, D.C.: Brookings Institution, 1977.

KAUFMAN, HERBERT. "The Natural History of Organizations." *Administration and Society* 7 (August 1975, as corrected at 365 in November 1975): 131-49.

KAUFMAN, HERBERT. "Why Organizations Behave as They Do: An Outline of a Theory." In *Papers Presented at an Interdisciplinary Seminar on Administrative Theory,* 37-72. Austin: University of Texas, 1961).

KEELEY, MICHAEL. "Organizational Analogy: A Comparison of Organismic and Social Contract Models." *Administrative Science Quarterly* 25, no. 2 (June 1980): 337-62.

KEMENY, JOHN G. "Man Viewed as a Machine." In *Automatic Control,* 132-46. New York: Scientific American Books, Simon and Schuster, 1955.

KIMBERLY, JOHN R.; MILES, ROBERT H.; and Associates. *The*

Organizational Life Cycle: Issues in the Creation, Transformation, and Decline of Organizations. San Francisco: Jossey-Bass, 1980.

LAWRENCE, PAUL R., and LORSCH, JAY W. *Organization and Environment.* Cambridge: Harvard University Press, 1967.

LEAVITT, HAROLD J. *Managerial Psychology: An Introduction to Individuals, Pairs and Groups in Organizations,* 2d ed. Chicago: University of Chicago Press, 1964.

LONG, NORTON E. "The Administrative Organization as a Political System." In *Concepts and Issues in Administrative Behavior,* edited by Sidney Mailick and Edward H. Van Ness, 110-21. Englewood Cliffs, N.J.: Prentice-Hall, 1962.

MACIVER, ROBERT M. *The Web of Government.* New York: Macmillan, 1947.

MARCH, JAMES G., and SIMON, HERBERT A. *Organizations.* New York: Wiley, 1958.

MARCH, JAMES G. "Bounded Rationality, Ambiguity, and the Engineering of Choice." *Bell Journal of Economics* 9, no. 2 (Autumn 1978): 587-608.

MARCH, JAMES G., ed. *Handbook of Organizations.* Skokie, Ill.: Rand McNally, 1965.

MARCHETTI, CESARE. "From the Primeval Soup to World Government: An Essay on Comparative Evolution." *Technological Forecasting and Social Change: An International Journal* 11, no. 1 (1977): 1-7.

MASTERS, ROGER D. "Genes, Language, and Evolution." *Semiotica* 2 (1970): 295-320.

MAYR, ERNST. *The Growth of Biological Thought: Diversity, Evolution, and Inheritance.* Cambridge: Harvard University Press, 1982.

MAYR, ERNST, and PROVINE, WILLIAM B., eds. *The Evolutionary Synthesis: Perspectives on the Unification of Biology.* Cambridge: Harvard University Press, 1980.

MERTON, ROBERT K. *Social Theory and Social Structure,* Revised and enlarged ed. New York: Free Press, 1957.

MEYER, MARSHALL W. *Environments and Organizations.* San Francisco: Jossey-Bass, 1978.

MIRVIS, PHILIP H., and BERG, DAVID N., eds. *Failures in Or-*

ganization Development and Change: Cases and Essays for Learning. New York: Wiley, 1977.

MUMFORD, LEWIS. The Myth of the Machine: Technics and Human Development. New York: Harcourt, Brace and World, 1967.

NELSON, RICHARD R., and WINTER, SIDNEY G. An Evolutionary Theory of Economic Change. Cambridge: Harvard University Press, 1982.

NYSTROM, PAUL C., and STARBUCK, WILLIAM H., eds. Handbook of Organizational Design, Vol. 1, Adapting Organizations to Their Environments; Vol. 2, Remodeling Organizations and Their Environments. New York: Oxford University Press, 1981.

ODELL, RICE. "Can Technology Avert the Errors of the Past?" Conservation Foundation Newsletter. Washington, D.C.: November 1979.

ODELL, RICE. "History Offers Some Warnings on Environment." Conservation Foundation Newsletter. Washington, D.C. October 1977.

OLSEN, MANCUR. The Rise and Decline of Nations: Economic Growth, Stagflation, and Social Rigidities. New Haven: Yale University Press, 1982.

OPARIN, A. I. The Origin of Life on the Earth. 3d ed. New York: Academic Press, 1957.

ORWELL, GEORGE. Nineteen Eighty Four. New York: Harcourt, Brace, 1949.

PATRICK, CUTHBERT T. The Lost Civilization: The Story of the Classic Maya. New York: Harper & Row, 1974.

PERROW, CHARLES. "Organizational Goals." International Encyclopedia of the Social Sciences. New York: Macmillan and Free Press, 1968.

PFEFFER, JEFFREY. Organizations and Organization Theory. Marshfield, Mass.: Pitman, 1982.

PFEFFER, JEFFREY. Power in Organizations. Marshfield, Mass.: Pitman, 1981.

PFEFFER, JEFFREY, and SALANCIK, GERALD R. The External Control of Organizations: A Resource Dependence Perspective. New York: Harper & Row, 1978.

POPPER, KARL R. *The Logic of Scientific Discovery.* New York: Basic Books, 1959.

PRESTHUS, ROBERT. *The Organizational Society: An Analysis and a Theory.* New York: Knopf, 1962.

REDFIELD, ROBERT, ed. *Levels of Integration in Biological and Social Systems.* Tempe, Ariz.: Jaques Cattell Press, 1942.

ROETHLISBERGER, F. J., and DICKSON, WILLIAM J. *Management and the Worker.* Cambridge: Harvard University Press, 1939.

ROURKE, FRANCIS E. *Bureaucracy, Politics, and Public Policy,* 2d ed. Boston: Little, Brown, 1976.

RUSH, J. H. *The Dawn of Life.* Garden City, N.Y.: Hanover House, 1957.

SAHLINS, MARSHALL D., and SERVICE, ELMAN R., eds. *Evolution and Culture.* Ann Arbor: University of Michigan Press, 1960.

SAYLES, LEONARD R., and CHANDLER, MARGARET K. *Managing Large Systems: Organizations for the Future.* New York: Harper & Row, 1971.

SCHUMPETER, JOSEPH A. *Capitalism, Socialism, and Democracy.* New York: Harper, 1942.

SCHUMPETER, JOSEPH A. *Theory of Economic Development: An Inquiry into Profits, Capital, Credit, Interest, and the Business Cycle.* Cambridge: Harvard University Press, 1934.

SCOTT, W. RICHARD. *Organizations: Rational, Natural, and Open Systems.* Englewood Cliffs, N.J.: Prentice-Hall, 1981.

SEIDENBERG, RODERICK. *Posthistoric Man: An Inquiry.* Chapel Hill: University of North Carolina Press, 1950.

SELZNICK, PHILIP. *Leadership in Administration: A Sociological Interpretation.* New York: Harper & Row, 1957.

SILLS, DAVID L. *The Volunteers: Means and Ends in a National Organization.* New York: Free Press, 1957.

SILLS, DAVID L. "Voluntary Associations: Sociological Aspects." *International Encyclopedia of the Social Sciences.* New York: Macmillan and Free Press, 1968.

SIMON, HERBERT A. *Administrative Behavior: A Study of Decision-making Processes in Administrative Organization.* New York: Macmillan, 1947.

SIMON, HERBERT A. "The Architecture of Complexity." *Proceedings of the American Philosophical Society* 106, no. 6 (December 1962): 470-73.

SIMON, HERBERT A. "On the Concept of Organization Goal." *Administrative Science Quarterly* 9 (1964): 1-22.

SIMON, HERBERT A.; SMITHBURG, DONALD W.; and THOMPSON, VICTOR A. *Public Administration.* New York: Knopf, 1950.

SIMPSON, GEORGE GAYLORD. *The Meaning of Evolution: A Study of the History of Life and of its Significance for Man.* New Haven: Yale University Press, 1949.

SPRINGLE, J. W. S. "The Origin of Life." Society for Experimental Biology. *Symposia* 7 (1953): 1-21.

STEBBINS, G. LEDYARD. *Darwin to DNA, Molecules to Humanity.* New York: Freeman, 1982.

STINCHCOMBE, ARTHUR L. "Social Structure and Organizations." In *Handbook of Organizations,* edited by James G. March, 142-93. Skokie, Ill.: Rand McNally, 1965.

TAYLOR, DONALD W. "Decision Making and Problem Solving." In *Handbook of Organizations,* edited by James G. March. 48-86. Skokie, Ill.: Rand McNally, 1965.

THOMPSON, VICTOR A. *Bureaucracy and Innovation.* University, Ala.: University of Alabama Press, 1969.

THOMPSON, VICTOR A. *Modern Organizations: A General Theory.* New York: Knopf, 1961.

TULLOCK, GORDON. *The Politics of Bureaucracy.* Washington, D.C.: Public Affairs Press, 1965.

URWICK, L. "Organization as a Technical Problem." In *Papers on the Science of Administration,* edited by Luther Gulick and L. Urwick, 49-88. New York: Institute of Public Administration, 1937.

WALKER, JACK L. "The Origins and Maintenance of Interest Groups in America." *American Political Science Review* 77, no. 2 (June 1983): 390-406.

WARSH, DAVID. *The Idea of Economic Complexity.* New York: Viking, 1984.

WILENSKY, HAROLD L. *Organizational Intelligence: Knowledge and Policy in Government and Industry.* New York: Basic Books, 1967.

WILSON, EDWARD O. *Sociobiology: The New Synthesis.* Cambridge: Harvard University Press, 1975.

WILSON, JAMES Q. *Political Organizations.* New York: Basic Books, 1973.

YUCHTMAN, EPHRAIM, and SEASHORE, STANLEY E. "A System Resource Approach to Organizational Effectiveness." *American Sociological Review* 32, no. 6 (December 1967).

ZALTMAN, GERALD, and DUNCAN, ROBERT. *Strategies for Planned Change.* New York: Wiley, 1977.

ZALTMAN, GERALD; DUNCAN, ROBERT; and HOLBEK, JONNY. *Innovations and Organizations.* New York: Wiley, 1973.

Name Index

171

Subject Index

177

About the Author

HERBERT KAUFMAN received his bachelor's degree from the City College of New York and his graduate degrees from Columbia University. He was for many years a member of the Department of Political Science at Yale University and then of the Governmental Studies Program at the Brookings Institution. A perennial student of organizations, his books include *The Forest Ranger: A Study in Administrative Behavior; The Limits of Organizational Change; Are Government Organizations Immortal?; Red Tape: Its Origins, Uses, and Abuses;* and *The Administrative Behavior of Federal Bureau Chiefs.*